THE STABILITY OF
ROTATING LIQUID MASSES

THE STABILITY OF
ROTATING LIQUID MASSES

BY

R. A. LYTTLETON

Fellow of St. John's College, Cambridge

CAMBRIDGE
AT THE UNIVERSITY PRESS
1953

CAMBRIDGE UNIVERSITY PRESS
Cambridge, New York, Melbourne, Madrid, Cape Town,
Singapore, São Paulo, Delhi, Mexico City

Cambridge University Press
The Edinburgh Building, Cambridge CB2 8RU, UK

Published in the United States of America by Cambridge University Press, New York

www.cambridge.org
Information on this title: www.cambridge.org/9781107615588

First published 1953
First paperback edition 2013

A catalogue record for this publication is available from the British Library

ISBN 978-1-107-61558-8 Paperback

CONTENTS

PREFACE

The present volume covers those parts of the theory of the stability of rotating gravitating liquids that seem to be of primary importance in determining the evolution of such systems. Apart from the intrinsic mathematical and dynamical interest of the subject, the problem is also of considerable interest from a cosmogonical standpoint, as its solution is the sole source of theoretical information on the question of how an isolated unstable rotating mass will develop. The important conclusion is reached, contrary to Jeans's views and to those still largely current amongst astronomers, that the dynamical evidence is entirely adverse to the so-called fission process of formation of binary systems. The work accordingly removes theoretical foundation from this process as playing any role in the evolution of binary systems. In this way the study indirectly assumes astrophysical value in that in disposing of the fission hypothesis it lifts what seems to the writer to have been one of the major obstacles to progress with the greater problem of stellar evolution.

My own interest in the subject of rotating gravitating liquids, as best I can remember, began twenty years ago with Professor H. F. Baker's lectures at Cambridge, when he used to conclude his course on celestial mechanics with a rapid survey of those parts of the present subject that could be dealt with by elementary methods. The astronomical importance of the problem became clear to me from studies of the origin of the solar system, and in particular it was from the problem of the origin of satellites that I was first led to suspect the validity of the fission process—this, incidentally, going to show how even conjectural studies can at times contribute suggestively to more precise matters. Investigation of the details of Jeans's researches soon disclosed numerous places where he had fallen into error, and I accordingly resolved to see if after rectifying these the subject could be brought into some more coherent form, not in conflict with the strong indications already available from other related fields. The requisite study was, however, far too protracted to be undertaken in any urgent or immediate way, and spread over several years it has had to take second and lower place to many other items that have occupied me, both of a scientific and non-scientific character, during much of that time. The present book is the outcome of this work, and while it deals for the most part only with a single line of attack aimed at settling the main cosmogonical question—that of the fission process—it is hoped that the discussion may serve to sum up and clarify progress to date and at the same time provide an introduction to the subject that may give at least initial help to others wishing to proceed to more recondite matters in which there is so much scope for further research.

Numerous references are supplied in the Appendix, and although these do not claim to be comprehensive there should be no difficulty by means of them in tracing almost any item of the literature of the subject.

St. John's College
Cambridge
April 1951

R. A. LYTTLETON

Chapter I

INTRODUCTION

The subject of the forms of relative equilibrium of a rotating mass of homogeneous gravitating liquid had its inception with the discussion by Newton (1687) of the figure of the Earth. In this it was simply assumed that a possible figure of the free surface would be that of an oblate spheroid with its least axis coincident with the axis of rotation, and it was not until Clairaut's work many years later that the validity of this postulate was examined. In the first instance Clairaut gave a proof resting on an approximate expression for the potential of a spheroid, but meanwhile Maclaurin (1740) produced an accurate demonstration of the possibility of the spheroidal form, and this led Clairaut also to publish an exact solution in place of his former one. It was rigorously shown by these authors that a spheroid is a possible equilibrium form whatever its eccentricity of meridian section provided it possesses an appropriate quantity of angular momentum.

That an ellipsoid with three unequal axes, the least coinciding with the axis of rotation, is also a possible form of relative equilibrium, provided the angular momentum is greater than a certain amount, remained undiscovered until Jacobi (1834) pointed it out in a letter to the French Academy. Jacobi himself does not appear to have published the result, and it seems first to have been referred to publicly by Poisson shortly after Jacobi's communication to the Academy. There is perhaps something of an element of surprise about Jacobi's result in view of the symmetry that might be expected to be associated with any form produced by a rotational field, and the fact also that the Jacobian figures exist only if the angular momentum exceeds a certain amount no doubt contributed to the series being overlooked for so long.

The first member of this Jacobian series is also a Maclaurin spheroid, but thereafter, for greater angular momentum, the equatorial axes are always different, and the elongation of the figure continually increases with the angular momentum. The limiting final form on this series, as infinite angular momentum is approached, has infinite longest axis, while the axis of intermediate length tends to equality with the third and least axis, both of them approaching zero.

Up to this stage researches had not gone beyond the question of the existence of possible equilibrium forms, nor did they do so until the problem was taken up by Poincaré (1885) in the paper that has since become celebrated. His investigation developed a method for studying the difficult question of the stability of the spheroidal and ellipsoidal forms, and necessarily involved also the consideration of the existence and properties of other equilibrium forms. There seems little doubt that a thorough discussion of these questions can be made only by using ellipsoidal harmonic analysis, which by the time Poincaré commenced his work had already been extensively developed by Lamé and others, and lay ready to hand.

As is well known, where rotating systems are concerned, the stability of steady configurations of relative equilibrium is a more complicated question than for statical systems, and in fact two different kinds of stability may be involved, usually designated by the terms 'secular' and 'ordinary'. The former contemplates the presence within the system of friction (vanishing with the relative velocities), whereas the latter is independent of dissipative action. Poincaré was able to show that if the mass of liquid is regarded as passing slowly along the Maclaurin series in the direction of increasing angular momentum, it becomes secularly unstable at a certain degree of flattening. This particular configuration, called a 'form of bifurcation', coincides with the beginning of the Jacobi series. The instability enters for a deformation of the surface involving a certain second order harmonic function. Moreover the Jacobi series can be regarded in its initial stages as the result of making a second-order harmonic deformation of the critical member of the Maclaurin series. This second-order deformation may be pictured as the addition of a stationary wave on the surface of the spheroid. This wave has two places of greatest elevation and two of least elevation on the equator of the spheroid, while the displacement vanishes at the poles, and it has the effect of transforming the spheroid into an ellipsoid with unequal axes. Thus the Jacobi series may be regarded as branching off the Maclaurin series and coming into existence as a result of a surface deformation involving precisely the harmonic terms through which the Maclaurin series has become unstable.

For greater values of the angular momentum along the Maclaurin series, whose members are all possible equilibrium forms, Poincaré showed that further separate points of onset of secular instability occur involving the higher order harmonic deformations. The proof of these results for the spheroids depends for the most part on properties of the ordinary zonal and tesseral harmonic functions to which the ellipsoidal harmonics reduce, in terms of suitable coordinates, when the ellipsoid has two axes equal. The corresponding investigation for the series of Jacobi forms involves the general ellipsoidal harmonics, but in a closely analogous way Poincaré was able to show that the Jacobi figures first become secularly unstable for a certain third-order harmonic deformation, and that for greater elongations and angular momentum, further separate configurations of instability occur, corresponding successively to harmonics of fourth-order, fifth-order, and so on.

The occurrence of these instabilities accordingly suggests the possibility that at each point of bifurcation there branches off a new series of configurations of equilibrium given initially in each case by the addition of a surface deformation expressed by the same harmonic through which the instability first appears. As the Jacobi series is described in the direction of increasing angular momentum, instability first enters for a certain third-order harmonic, and the initial forms of the branch series will thus be obtainable by adding a surface deformation of this kind. The surface wave involved in this case has three places of greatest elevation and three of least elevation in the equatorial plane, and vanishes at all points in the plane orthogonal to this defined by the two shorter axes. The amplitudes of the deformations supposed added are assumed infinitesimal throughout, but if a small finite amplitude is adopted for the purposes of illustrating the nature of the new equilibrium forms, the resulting surface (Fig. 16, p. 110) was considered by Poincaré to bear marked resemblance to that of a pear. Accordingly the initial figures of the series became known as the 'pear-shaped' figures, or sometimes the 'piriform' figures. The difficult problem that accordingly next arose was to ascertain whether or not the pear-shaped series is initially stable or unstable. This in essence is Tchebychef's problem, and it gave rise to extensive investigations by Poincaré, Darwin, Liapounoff, and Jeans, to mention the chief authors. Minor contributions on various technical points connected with it were made by Schwarzschild, H. F. Baker, and others.

The properties of the fluid mass having been assigned, as uniform, gravitating, and if necessary viscous, the general problem of possible equilibrium forms and their stability can be stated as a purely theoretical one, but Poincaré to some extent, and Darwin almost entirely, were interested in the problem from its possible cosmogonical applications. The general form of the pear-shaped figure undoubtedly gave rise to the notion that if the mass were stable and evolved by equilibrium forms along this series, with the furrow continually deepening as the figure elongated, the final result would be two detached masses rotating in circular orbital motion about each other. Thus it seemed plain to Darwin that the dynamical theory, if it could be established in accordance with these ideas, would give a strong theoretical basis for supposing this to be the method of genesis of double systems in the celestial universe. Indeed, Darwin eventually announced that he had proved the pear-shaped series to be initially stable and hence that this would in fact be the course of development.

Now in order to establish that the pear-shaped figures are stable it is sufficient to show that they are secularly stable. On the other hand, if the series were shown to be secularly unstable, it would in general require further investigation to decide in what manner the system is next likely to develop. At all events, Darwin approached the problem by a method that aimed only at settling the question of secular stability, as also did Liapounoff, though this author appears to have been attracted solely by the theoretical problem, and in no wise interested in the astronomical implications of the results, which pass unmentioned in his numerous papers. Subsequently, Jeans also concerned himself with establishing the secular stability or otherwise, apparently under the impression that ordinary stability was not relevant to the problem.

Had these writers been able finally to conclude that the pear-shaped figure was secularly stable, their treatments would have represented a complete solution of the immediate problem (though there would have still remained the question of how far the pear-shaped series continued to be stable with increasing angular momentum, just as this was necessary for the Maclaurin and Jacobi series). But in point of fact Liapounoff, and later Jeans, concluded that the figure was quite certainly secularly unstable, and moreover Jeans claimed that Darwin's original investigation, when properly interpreted with certain technical errors set right, itself in reality led to this same result. As far as the objective of Liapounoff's inquiry was concerned he had attained it, but Jeans's expressed intention was to obtain theoretical evidence of the course of dynamical stellar evolution, albeit under what have since proved perhaps unduly idealized conditions. The desired information, however, is not conclusively provided by the knowledge that the system is secularly unstable, for this means that the pear-shaped form itself never comes into existence, and if the continuation of the Jacobi series remained ordinarily stable, as might possibly happen, the rate of departure from the critical Jacobi figure might not take place at all rapidly if frictional effects were small. For example, the lunar orbit is secularly unstable, but ordinarily stable, and its evolution under frictional forces proceeds extremely slowly and would cease altogether if friction were absent.

Thus to complete the information that may be derived from the consideration of small deformations of the system it is necessary to determine whether the Jacobi series remains *ordinarily* stable or not beyond the critical Jacobi figure. It is automatically so before this stage is reached as a consequence of its secular stability. This question requires a totally different treatment from that of secular stability, since it is necessary for its solution to study the actual periods of possible small oscillations of the system, and not merely the manner in which some single entity, such as the moment of momentum, changes along the initial stages of the pear-shaped series. The determination of the ordinary stability of the Jacobi series has been undertaken and solved by Cartan, who has succeeded in proving that for displacements involving the third-order harmonic through which the ellipsoids first become secularly unstable, they simultaneously become ordinarily unstable.

With this information available the nature of the development beyond the critical Jacobi figure can be studied with more certainty. For if the free surface receives a displacement involving third-order harmonics, and any general physical disturbance may be assumed to contain such terms in its expression, the amplitudes will begin to increase exponentially at a rate independent of the amount of friction present. The system can no longer oscillate about an equilibrium form, since none exists of a stable character, and instead a dynamical motion must ensue until the system succeeds in finding its way to a new state of steady motion. The equations of small motion of the system permit this development to be followed only so long as the velocities and displacements involved in it remain small, but with increasing amplitude the approximations leading to linear equations of motion become less accurate. The system must, however, reach eventually some other steady condition involving no further dissipation of energy, and the interesting question arises as to what this final configuration will be. Unfortunately it is not possible to investigate the question in detail by anything approaching rigorous means, but it may well be, as has always been maintained, that the result will be a division of the original mass into two detached portions. However, if this view is correct, there is necessarily an important difference from Darwin's ideas on the course of development, for it can be shown not only that the pieces must be of considerably different sizes, but what is more important still, that they must separate to infinity. The final steady state would then consist of two separate unequal stable masses receding with constant relative velocity, the original excess angular momentum causing the instability reappearing now as orbital angular momentum.

From the point of view of cosmogony the main question of interest is to obtain as rigorous a demonstration of this course of development as possible. It would also be of interest to give as complete an account as possible of the whole evolution of the problem, but the literature of the subject is so extensive and much of it of an exploratory character, that it would scarcely be practicable in a single volume to give more than an outline of it. Yet it has seemed worthwhile to give the full mathematical discussion of such parts as are essential to establishing the extreme plausibility of the course of evolution described above. To do this we begin by discussing the subject of stability with particular reference to rotating systems. This is followed by a discussion of the spherical, spheroidal, and ellipsoidal forms, together with certain of their properties that can be established by simple means, as illustrations of the dynamical theory. Next is developed the ellipsoidal harmonic analysis required for further progress, when the necessary properties of Lamé's functions are derived, and then using this mathematical technique an account is given of Poincaré's investigation of the secular stability of the Maclaurin and Jacobi series. This is followed by an account of Cartan's discussion of the ordinary stability of the Jacobi forms. Finally, the subsequent development of the system is considered and its possible cosmogonical implications discussed.

Chapter II

STABILITY

STABILITY OF STATICAL SYSTEMS

Equilibrium configurations

Let us consider a mechanical system whose position can be specified by n generalized coordinates $q_1, q_2, ..., q_n$, and its motion at any instant by the generalized velocities $\dot{q}_1, \dot{q}_2, ..., \dot{q}_n$, where the dots denote differentiation with respect to time t. Also let us suppose that the forces acting on the system are derivable from a potential energy function V depending on the coordinates only, so that

$$V = V(q_i).$$

The kinetic energy T will in general depend on both the coordinates and the velocities, and will be a homogeneous quadratic function of the latter. Thus

$$T = T(q_i, \dot{q}_i).$$

Motion of the system will take place in accordance with the Lagrangian equations

$$\frac{d}{dt}\left(\frac{\partial T}{\partial \dot{q}_i}\right) - \frac{\partial T}{\partial q_i} = -\frac{\partial V}{\partial q_i} \quad (i = 1, 2, ..., n), \tag{1}$$

which possess the energy integral

$$T + V = \text{constant.} \tag{2}$$

Hence the possible equilibrium configurations, within the limits of the number of coordinates adopted, are determined by the n equations

$$\frac{\partial V}{\partial q_i} = 0 \quad (i = 1, 2, ..., n), \tag{3}$$

which are simply the conditions that V is stationary.

These equations may have certain admissible solutions, each of which will correspond to a possible equilibrium state, and we may denote a particular one by

$$q_i = a_i \quad (i = 1, 2, ..., n). \tag{4}$$

Moreover, it can readily be seen that if V is an absolute minimum in the configuration then the equilibrium is stable. For, in accordance with the energy integral (2), if a slight disturbance from equilibrium occurs, it follows, since T cannot become negative, that V cannot increase above its equilibrium value by more than a very small amount. This means in turn that none of the coordinates can deviate by more than a small amount from its equilibrium value, and therefore that during the motion the system must always remain in the immediate neighbourhood of the equilibrium configuration, which is what is meant by stability.

Linear series of configurations

If the description of the system contains a parameter μ, say, not itself dependent on any of the generalized coordinates, which for any reason is slowly changing, or is so regarded, the potential V will in general depend on μ, and the values of q_i giving the solutions of (3) will also depend on μ. Thus the particular solution (4) may be written

$$q_i = a_i(\mu) \quad (i = 1, 2, ..., n),$$

where the a_i's are now functions of μ.

When μ undergoes a small change $d\mu$, the solution will become

$$q_i = a_i(\mu + d\mu) = a_i + \frac{da_i}{d\mu} \cdot d\mu, \tag{5}$$

and hence the original configuration will give place to an adjacent configuration for the new system with slightly different μ. Thus starting from a given equilibrium position, a continuous series of other possible configurations is obtained as μ slowly varies. Such a set of configurations Poincaré has termed a 'linear series'. The essence of the idea is the presence within the system of a changing parameter which, however, varies sufficiently gradually for the change not to affect equilibrium at any stage.

Stability

Let us consider next what may happen to the stability of the system as it moves slowly along a linear series. We examine this question first by diagrammatic means. If we take $(q_i; \mu)$ as rectangular coordinates with the μ-axis vertical, then for different values of V on the right-hand side, the relation

$$V(q_i; \mu) = V \tag{6}$$

will correspond in two dimensions to a family of curves, or in three dimensions to surfaces. With more than two coordinates, q_i, the relation will represent hyper-surfaces according to the number of dimensions involved. As there is only one value of V for a given point $(q_i; \mu)$ these surfaces are essentially non-intersecting.

The condition that the tangent plane to a given V-surface shall be perpendicular to the μ-axis is simply that $dV = 0$ for $d\mu = 0$, which is equivalent to the relations (3). The equilibrium configurations therefore correspond to the points where the V-surfaces are horizontal. In Fig. 1, let V_1 and V_2 represent two surfaces of the family and suppose $V_2 > V_1$. Then the point P_1 represents an equilibrium configuration and the corresponding value of μ is μ_1, as shown. The heavy line joining successive equilibrium points $P_1, P_2, \ldots$ represents the linear series.

Slight displacements of the system, without change of μ, will be represented, for $\mu = \mu_1$ say, by points such as P_1' in the tangent plane at P_1. In the circumstances postulated, such changes will involve an increase of potential energy if the V-surfaces are concave downwards as shown, since the general direction of V increasing is as indicated. Thus the condition for stability can be stated in the form that the V-surfaces must be concave downwards.

Clearly, if the general direction of increasing V were in the present case the other way, stability would require the surfaces to be concave upwards. The rule is therefore that for stability the concavities must be towards the direction of V decreasing.

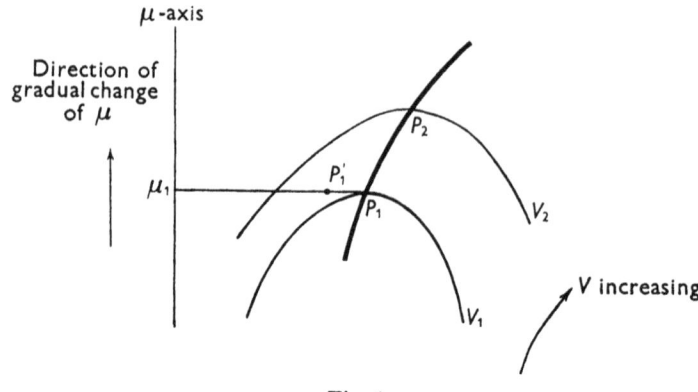

Fig. 1.

Exchange of stabilities

As the representative point of the configuration passes along the linear series no new feature arises so long as the V-surfaces remain concave in the same direction. However, it may happen that they gradually develop in one of the ways shown in the diagrams of Fig. 2, which illustrate some of the possible forms that the V-surfaces may take. The analytical discussion given later covers all possible cases. At present we consider these four examples.

(i) As the linear series is described in the direction indicated by the arrow, the concavities eventually become the other way and additional downward concavities appear. Accordingly, at the critical point C another linear series BCB cuts across the original series $A_1 CA_2$. Such a point is termed a 'point of bifurcation', and the corresponding equilibrium position a 'form of bifurcation'. If the series $A_1 C$ was at first stable, the change in direction of the concavities means that beyond C the continuation CA_2 of this series must be unstable. That is, at C the original series loses its stability. On the other hand, in the case depicted, the new series BCB has its concavities the same way as the series $A_1 C$, and hence the configurations of this series will be stable. There has occurred a transfer of stability to the members of the new series. For each value of μ corresponding to the portion $A_1 C$, there exists only one equilibrium form, and this is stable, but for each value beyond this there are three possible forms of which two are stable and one unstable.

(ii) Here, assuming the initial series $A_1 C$ to be stable, the continuation CA_2 of this series must again be unstable, but the series BCB must now also be unstable. In this case there is therefore a disappearance of stability at C.

(iii) Here two configurations of equilibrium are possible to begin with for each value of μ, and if that on the series $A_1 C$ is stable, that on CA_2 is unstable. Not only is stability lost at C, but for values of μ beyond that of C there are no equilibrium forms possible at all.

(iv) Here only one configuration is possible for values of μ corresponding to the series $A_1 C$, and again there are no equilibrium forms possible beyond that corresponding to C.

Cases (i) and (ii) are of special importance in the actual problems with which we shall be concerned. In the application of these considerations to physical systems it is convenient always to choose for μ some parameter that increases,

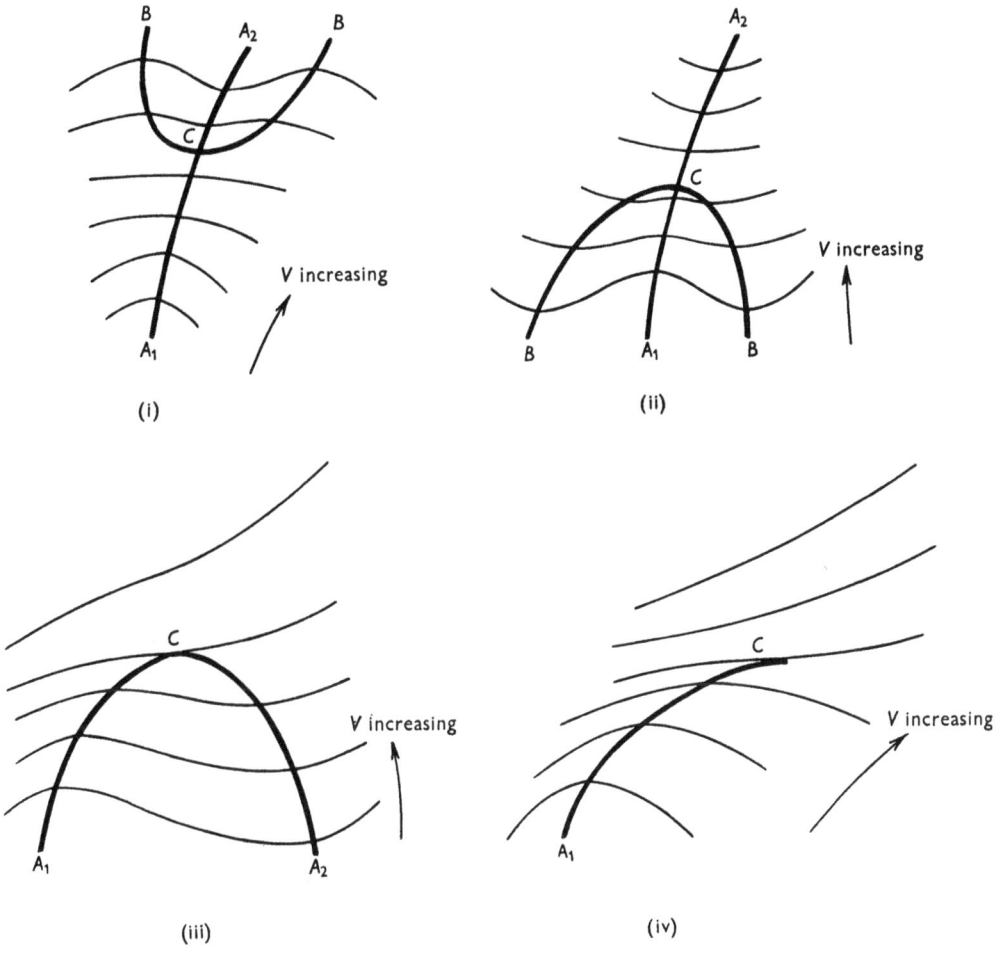

Fig. 2.

or decreases, monotonically as the system gradually evolves. We shall see that one or other of such quantities as the angular velocity or angular momentum can usually be postulated to satisfy this requirement. Assuming such a choice to have been made, it is seen that in (ii) the members of the unstable series BCB cannot actually come into existence, even though the value of μ would permit it theoretically, if the system is originally in a stable configuration on $A_1 C$. It could, of course, if necessary, be regarded as placed in one of these unstable positions.

It may be remarked that the term 'exchange of stabilities', which is often used in connexion with cases such as (i) and (ii), is to some extent a misleading one, because stability is not necessarily transferred to the new series. There is always a loss of stability of the original series, but the new series, if one exists, may or may not be stable.

Following Jeans, these possible cases can be represented schematically by the following idealized diagrams:

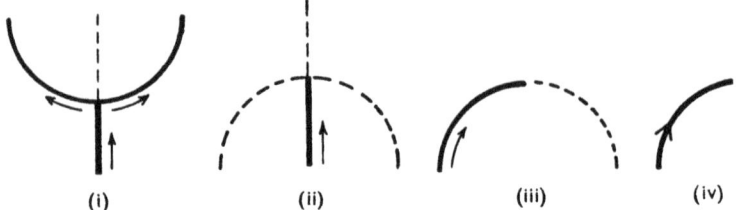

(i)   (ii)   (iii)   (iv)

Fig. 3. The heavy lines denote stable series and the dotted lines unstable series. The arrows indicate the direction of evolution.

Rules for graphical determination of stability

These considerations therefore lead to the following rules for determining diagrammatically how the stability of a system changes at a point of bifurcation:

If in evolving along a stable series represented by a vertical line a point of bifurcation is reached, stability will be lost by the first series and transferred to the new series if the curve representing the latter turns upwards. This is case (i) above.

If the new series turns downwards, as in (ii), configurations on it cannot come into existence, and the continuation of the original series is unstable.

If the original series itself turns downwards, as in (iii), or ceases altogether, as in (iv), there are no further equilibrium forms either stable or unstable for values of the parameter immediately beyond that corresponding to the point of bifurcation.

It is highly important to understand that the number of coordinates used to describe the system must remain the same throughout for these results to be valid. Continuous systems, such as liquid masses, require for their complete description an infinite number of coordinates, but by restricting the mass to special forms the number of coordinates required can be made finite. Thus, if only ellipsoidal forms are admitted, then but two coordinates are needed, for if a, b, c denote the semi-axes these must always be related by $abc = $ constant. If only spheroidal forms are admitted, then $a = b$ also, and one coordinate suffices. If a diagram of the above kind is drawn for the Maclaurin series no point of bifurcation appears because no other equilibrium series is possible within these restrictions. On the other hand, if a diagram is drawn for the ellipsoidal series, it is found to have a point of bifurcation at the point where the Maclaurin series crosses it.

It is for the present reason that the possibility of the pear-shaped figure only became demonstrated when the full number of freedoms could be coped with by the method of ellipsoidal harmonics, though something of the kind had already been vaguely conjectured by Kelvin and Darwin.

Analytical discussion

The condition for a point of bifurcation on a linear series can readily be determined analytically.

Supposing $q_i = a_i$ is an equilibrium configuration when the value of the parameter is μ, so that $\dfrac{\partial V}{\partial q_i} = 0$ for this configuration, let us consider the value of the potential energy $V(q_i; \mu)$ in an adjacent configuration given by $(a_i + \delta q_i; \mu + \delta\mu)$. As far as second-order terms in the small quantities $\delta q_i, \delta\mu$, we have

$$V = V(q_i; \mu) + \frac{1}{2}\frac{\partial^2 V}{\partial q_r \partial q_s}\delta q_r \delta q_s + \frac{\partial V}{\partial \mu}\delta\mu + \frac{1}{2}\frac{\partial^2 V}{\partial \mu^2}\delta\mu^2 + \frac{\partial^2 V}{\partial \mu \partial q_r}\delta q_r \delta\mu, \qquad (7)$$

where summation is made over repeated suffixes.

For a fixed value of $\delta\mu$, the quantities δq_i may be regarded as the coordinates of the system, and the conditions that $(a_i + \delta q_i; \mu + \delta\mu)$ represents a new equilibrium configuration are the n equations

$$\frac{\partial V}{\partial(\delta q_i)} = 0 \quad (i = 1, 2, \ldots, n).$$

Writing V_{rs} for $\partial^2 V/\partial q_r \partial q_s$ and $V_{r\mu}$ for $\partial^2 V/\partial q_r \partial\mu$, these conditions, written at length, have the form:

$$\left.\begin{array}{l} V_{11}\delta q_1 + V_{12}\delta q_2 + \ldots + V_{1n}\delta q_n + V_{1\mu}\delta\mu = 0 \\ V_{21}\delta q_1 + V_{22}\delta q_2 + \ldots + V_{2n}\delta q_n + V_{2\mu}\delta\mu = 0 \\ \ldots \quad\quad \ldots \quad\quad\quad \ldots \quad\quad \ldots \\ V_{n1}\delta q_1 + V_{n2}\delta q_2 + \ldots + V_{nn}\delta q_n + V_{n\mu}\delta\mu = 0. \end{array}\right\} \qquad (8)$$

The solution of these may be written in determinantal form as follows:

$$\frac{\delta q_1}{\begin{vmatrix} V_{12} & V_{13} & \cdots & V_{1n} & V_{1\mu} \\ \cdots & \cdots & \cdots & \cdots & \cdots \\ V_{n2} & \cdots & \cdots & \cdots & \cdots \end{vmatrix}} = \frac{\delta q_2}{\begin{vmatrix} V_{11} & V_{13} & \cdots & \cdots & V_{1\mu} \\ \cdots & \cdots & \cdots & \cdots & \cdots \\ V_{n1} & \cdots & \cdots & \cdots & \cdots \end{vmatrix}} = \ldots = \frac{\delta\mu}{\Delta}, \qquad (9)$$

where

$$\Delta = \begin{vmatrix} V_{11} & V_{12} & \cdots & \cdots & V_{1n} \\ V_{21} & V_{22} & \cdots & \cdots & V_{2n} \\ \cdots & \cdots & \cdots & \cdots & \cdots \\ V_{n1} & V_{n2} & \cdots & \cdots & V_{nn} \end{vmatrix}.$$

These equations determine the requisite step δq_i (in a space of $n+1$ dimensions) from a known equilibrium configuration a_i to a new equilibrium configuration for a given increment $\delta\mu$ of the parameter μ. They will fail to determine such a step only if

$$\Delta = 0, \qquad (10)$$

and this is the analytical condition for a point of bifurcation or a point where a series ceases with increasing μ.

Returning to the diagrams of Fig. 2 for a moment, we see that in case (i) if the original configuration is just below C, there are three possible adjacent equilibrium configurations for a small increment $\delta\mu$, so that for such a system Δ must be zero at C, as otherwise a unique adjacent configuration would necessarily exist. Similarly in case (ii) Δ must vanish at C. In cases (iii) and (iv) the linear series are such that $\delta\mu$ must itself be zero if the configuration is to remain on the series, and hence in these cases too Δ must be zero at C.

Whenever this happens the next stage is indeterminate to the present order of approximation and could therefore be found only by proceeding to a higher order, when evidently the equations corresponding to (8) would no longer be linear, and their solution would give more than one set of values for the δq_i. These solutions would presumably be real for positive $\delta\mu$ in a case such as (i), and for a negative $\delta\mu$ in a case such as (ii), whereas for (iii) and (iv) their solution would be imaginary and therefore not correspond to an admissible step to a real adjacent configuration.

Stability

Let us examine next how the vanishing of Δ affects the stability of the original series of configurations $q_i = a_i$.

The parameter μ is now to be kept constant, so that the variable part, δV say, of the potential energy for adjacent configurations is

$$\delta V = \tfrac{1}{2} V_{rs} \delta q_r \delta q_s.$$

By means of a real non-singular linear transformation of the coordinates, now δq_i, this expression can be transformed to a sum of square terms only, and denoting by $\theta_1, \theta_2, ..., \theta_n$ a new set of coordinates chosen in this way, we shall have

$$\delta V = \tfrac{1}{2}(b_1 \theta_1^2 + b_2 \theta_2^2 + ... + b_n \theta_n^2),$$

wherein $b_1, b_2, ..., b_n$ are certain constants depending on the original position of the system and on μ. Let us denote the determinant of this transformation by m, so that m will be essentially non-zero. By a known theorem, the discriminant of the new quadratic form will be simply m times that of the original form. Thus we must have

$$
\begin{vmatrix}
b_1 & 0 & 0 & \dots & 0 \\
0 & b_2 & 0 & \dots & \dots \\
0 & 0 & b_3 & \dots & \dots \\
\dots & \dots & \dots & \dots & \dots \\
\dots & \dots & \dots & \dots & b_n
\end{vmatrix}
= m
\begin{vmatrix}
V_{11} & V_{12} & \dots & \dots & V_{1n} \\
V_{21} & V_{22} & \dots & \dots & V_{2n} \\
\dots & \dots & \dots & \dots & \dots \\
\dots & \dots & \dots & \dots & \dots \\
V_{n1} & V_{n2} & \dots & \dots & V_{nn}
\end{vmatrix}
$$

which reduces to $$b_1 b_2 b_3 ... b_n = m\Delta. \tag{11}$$

For the original configuration to be stable, the potential energy δV must be positive definite, and, by a known theorem, if this is so it means that $b_1, b_2, ..., b_n$ must all be positive. Whatever their values, positive or negative, the quantities $b_1, b_2, ..., b_n$ are called the 'coefficients of stability' of the system. They are not unique, since a quadratic form can be transformed to square terms only in an infinite number of ways; but however the reduction is made, the numbers of positive and negative coefficients in each of the resulting expressions are invariable.

If the original system is stable, $b_1, b_2, ..., b_n$ will all be positive initially, and a change from stability to instability will occur, as μ changes, only when one of these coefficients vanishes and changes sign. The values of μ for which this occurs are accordingly given by the same equation as that determining the points of bifurcation, namely, $\qquad \Delta = 0.$

It may therefore be concluded that at a point of bifurcation there occurs a loss or gain of stability. If the system is describing a part of a linear series that is stable, it must lose its stability for the configurations on the continuation of this series. If a branch series passes through the point for which $\Delta = 0$, it will be stable if its representative curve turns upwards so that new equilibrium configurations on it become possible as the parameter μ continues to increase. If it turns downwards these configurations will be unstable and cannot come into existence.

We give next some simple examples of statical systems illustrating the foregoing theory.

Examples of Stability of Equilibrium of Statical Systems

(i) *A heavy uniform rod AB of mass M and length 2a is supported by two equal crossed strings BC, AD of lengths 2b attached to two points C and D at distance 2a apart at the same level, and the system is constrained to move in the vertical plane through CD.*

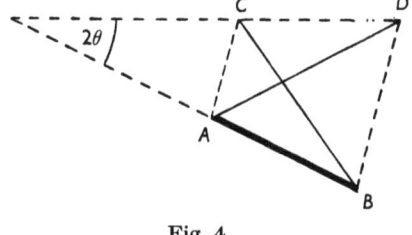

Denoting by 2θ the inclination of the rod to the horizontal, consideration of the depth of the mid-point of AB below the level CD gives immediately

$$U = V/2Mg = -\cos\theta\,(b^2 - a^2\cos^2\theta)^{\frac{1}{2}}$$

Fig. 4.

where g is the acceleration of gravity. Equilibrium positions are therefore given by

$$\frac{dU}{d\theta} = \frac{\sin\theta\,(b^2 - 2a^2\cos^2\theta)}{(b^2 - a^2\cos^2\theta)^{\frac{1}{2}}} = 0,$$

that is, $\qquad \sin\theta = 0, \quad \text{or} \quad \cos^2\theta = b^2/2a^2.$

The relevant solutions are

$$\theta = 0, \quad \text{or} \quad \theta = \pm\cos^{-1}(b/a\sqrt{2}).$$

The first corresponds to the horizontal position of the rod and is always a possible equilibrium configuration. The second, when it is a real angle, gives two symmetrical positions in which the rod is inclined to the horizontal.

To examine the stability of these various positions, we have

$$\frac{d^2 U}{d\theta^2} = \frac{\cos\theta\,(b^2 - 2a^2\cos^2\theta)}{(b^2 - a^2\cos^2\theta)^{\frac{1}{2}}} + \frac{a^2\sin^2\theta\cos\theta\,(3b^2 - 2a^2\cos^2\theta)}{(b^2 - a^2\cos^2\theta)^{3/2}}.$$

If $\theta = 0$, this gives $U_{\theta\theta} = (b^2 - 2a^2)/(b^2 - a^2)^{\frac{1}{2}}$ and the position is therefore stable if $b > a\sqrt{2}$, and unstable if $b < a\sqrt{2}$. Thus the second inclined positions only exist when the horizontal position is unstable. If $\cos^2\theta = b^2/2a^2$, we find

$$U_{\theta\theta} = 4a\left(1 - \frac{b^2}{2a^2}\right) = 4a\sin^2\theta,$$

and since this is essentially positive, these inclined positions, when they exist, are always stable.

If, therefore, the system is imagined to begin with a/b small, and gradually increasing, the series of equilibrium positions will provide an instance of case (i) of the foregoing theory. The horizontal position is the only equilibrium form possible, and is stable, until $a\sqrt{2}$ becomes greater than b, when two additional stable equilibrium forms become possible, while the horizontal position becomes unstable.

(ii) *A heavy bead of mass m can slide on a smooth vertical circular wire, radius a, and is attached to a second particle of mass M by an inextensible string that passes over a smooth peg at the highest point of the circle, the mass M hanging freely.*

Denoting by θ the angle between the two portions of the string, as shown, then we find

$$U = V/2ga = M\cos\theta - m\cos^2\theta.$$

Equilibrium positions are therefore given by

$$\frac{dU}{d\theta} = -M\sin\theta + 2m\cos\theta\sin\theta = 0,$$

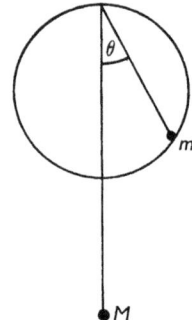

that is, $\theta = 0$,   or $\theta = \pm\cos^{-1} M/2m$,

which gives real θ only if $M \leqslant 2m$.
To consider stability, we have

$$\frac{d^2 U}{d\theta^2} = -M\cos\theta + 2m(2\cos^2\theta - 1).$$

Fig. 5.

If $\theta = 0$, $U_{\theta\theta} = -M + 2m$ and is positive if $M < 2m$.
If $\cos\theta = M/2m$, $U_{\theta\theta} = (M^2 - 4m^2)/2m$ and is negative if $M < 2m$.

If therefore M/m is taken as parameter and regarded as gradually increasing from zero, there are at first three possible equilibrium positions of which $\theta = 0$ is stable and the two symmetrically placed positions on either side of the vertical are unstable. For $M/m > 2$, only the former position is possible and it is then unstable. The system is therefore an example of case (ii) of the theory.

(iii) *A heavy bead can slide under gravity on a smooth wire in the form of a conic-section with major-axis horizontal and minor-axis vertical.*

For the parameter let us take the eccentricity e of the curve, and suppose that the latus rectum $2b^2/a$ remains fixed. Then for $e < 1$, equilibrium positions are at the ends of the minor axis. The lowest position is stable and the highest unstable. For $e > 1$, however, the curve becomes a hyperbola and no equilibrium positions exist. We have therefore an instance of case (iii).

(iv) *A heavy particle can slide under gravity on a smooth circular wire in a vertical plane and is attached by means of an elastic string to a point A of the circumference at angular distance α from the highest point.*

Supposing α is initially zero and the string sufficiently short for the particle to be in contact with the curve, then if α is gradually increased, so that the particle is drawn up the curve, there will be only one possible equilibrium position for each value of α, and this position will eventually reach the highest point. If any further increase in α occurs, the system will cease to be a statical one, so that the linear series will end abruptly.

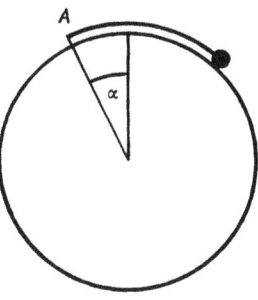

Fig. 6.

STABILITY OF ROTATING SYSTEMS

Where rotating systems are concerned, the configurations of importance are those of relative equilibrium in which the whole system rotates steadily about a fixed axis through the centre of mass as if it were a rigid body. In such a state there are no relative motions of the parts, so that as far as internal influences are concerned no dissipation of energy can occur and the system remains steady. If relative motions do occur for any reason there will in general no longer be a unique angular velocity of the system, though the direction of the angular momentum vector may provide a fixed direction at the centre of mass. If so, we can adopt a set of rectangular rotating axes at this point of which the third, say, is fixed and the other two rotate round it. The positions of the particles can then be referred to this rotating frame.

Let us suppose the axes Ox, Oy, Oz to be rectangular and rotating about Oz, defined in this way, with angular velocity $\omega = \dfrac{d\psi}{dt}$, not necessarily constant. If a particle of mass m has coordinates (x, y, z) in the rotating frame, its space velocity at any instant will be $(\dot{x} - \omega y, \dot{y} + \omega x, \dot{z})$. Hence the kinetic energy of a system of such particles is given by

$$T = \tfrac{1}{2}\Sigma m\{(\dot{x} - \omega y)^2 + (\dot{y} + \omega x)^2 + \dot{z}^2\}$$
$$= \tfrac{1}{2}\Sigma m(\dot{x}^2 + \dot{y}^2 + \dot{z}^2) + \omega\Sigma m(x\dot{y} - y\dot{x}) + \tfrac{1}{2}\omega^2\Sigma m(x^2 + y^2).$$

The angular momentum round the z-axis relative to a fixed frame of reference is given by

$$H = \Sigma m\{(\dot{y} + \omega x)x - (\dot{x} - \omega y)y\}$$
$$= \Sigma m(x\dot{y} - y\dot{x}) + \omega\Sigma m(x^2 + y^2).$$

If we write

$$T_R = \tfrac{1}{2}\Sigma m(\dot{x}^2 + \dot{y}^2 + \dot{z}^2),$$

$$H_R = \Sigma m(x\dot{y} - \dot{x}y),$$

$$I = \Sigma m(x^2 + y^2),$$

then T_R is the kinetic energy relative to the rotating frame, and H_R is the angular momentum relative to this frame. Also I is the moment of inertia of the system round the z-axis, and is independent of the frame in which the rectangular coordinates x, y happen to be measured. Expressing T and H in terms of these quantities we have

$$T = T_R + \omega H_R + \tfrac{1}{2}\omega^2 I, \qquad (12)$$

$$H = H_R + \omega I, \qquad (13)$$

and, by eliminating H_R, we have also

$$T = T_R + \omega H - \tfrac{1}{2}\omega^2 I. \qquad (14)$$

Let it be supposed next that the system is capable of being described by $n+1$ coordinates, consisting of n generalized coordinates q_i that fix the configuration relative to the rotating frame, and the azimuthal coordinate ψ, which will be assumed not to occur explicitly but only as $\dot{\psi}$. Assuming, as will be the case for natural systems, that the internal forces are derivable from a potential function V, and that the only external force is a couple G round the z-axis, the Lagrangian equations of motion for the system are

$$\frac{d}{dt}\left(\frac{\partial T}{\partial \dot{q}_i}\right) - \frac{\partial T}{\partial q_i} = -\frac{\partial V}{\partial q_i} \quad (i = 1, 2, ..., n), \qquad (15)$$

and

$$\frac{d}{dt}\left(\frac{\partial T}{\partial \omega}\right) - \frac{\partial T}{\partial \psi} = G. \qquad (16)$$

But from (12) $\dfrac{\partial T}{\partial \psi} = 0$ and $\dfrac{\partial T}{\partial \omega} = H$, so that (16) becomes

$$\frac{dH}{dt} = G, \qquad (17)$$

which expresses the equality of the rate of change of angular momentum round Oz with the external couple.

For a freely rotating system G will be zero and H constant. If, on the other hand, the system can always be described by means of n generalized coordinates relative to axes constrained to rotate uniformly with angular velocity ω, the time not occurring explicitly in the description, an external couple will be required of amount given by (16). For our purposes there are two main cases to consider. First, we shall take $\omega = $ constant, and second, the case $H = $ constant. We shall also show in what circumstances the two cases may be regarded as equivalent.

Systems rotating with constant angular velocity

To transform the Lagrangian equations (15), we treat the q_i's and $\dot{q}_i$'s as independent in the usual way, so that we have

$$\dot{x} = \frac{\partial x}{\partial q_i}\dot{q}_i, \qquad \frac{\partial \dot{x}}{\partial \dot{q}_i} = \frac{\partial x}{\partial q_i}, \qquad \text{and} \qquad \frac{\partial \dot{x}}{\partial q_i} = \frac{d}{dt}\frac{\partial x}{\partial q_i}.$$

Then

$$\frac{\partial H_R}{\partial \dot{q}_i} = \Sigma m\left(x\frac{\partial \dot{y}}{\partial \dot{q}_i} - y\frac{\partial \dot{x}}{\partial \dot{q}_i}\right) = \Sigma m\left(x\frac{\partial y}{\partial q_i} - y\frac{\partial x}{\partial q_i}\right),$$

so that

$$\frac{d}{dt}\left(\frac{\partial H_R}{\partial \dot{q}_i}\right) = \Sigma m\left(\dot{x}\frac{\partial y}{\partial q_i} - \dot{y}\frac{\partial x}{\partial q_i}\right) + \Sigma m\left(x\frac{\partial \dot{y}}{\partial q_i} - y\frac{\partial \dot{x}}{\partial q_i}\right),$$

and

$$\frac{\partial H_R}{\partial q_i} = \Sigma m\left(\dot{y}\frac{\partial x}{\partial q_i} - \dot{x}\frac{\partial y}{\partial q_i}\right) + \Sigma m\left(x\frac{\partial \dot{y}}{\partial q_i} - y\frac{\partial \dot{x}}{\partial q_i}\right).$$

Hence

$$\frac{d}{dt}\left(\frac{\partial H_R}{\partial \dot{q}_i}\right) - \frac{\partial H_R}{\partial q_i} = 2\Sigma m\left(\dot{x}\frac{\partial y}{\partial q_i} - \dot{y}\frac{\partial x}{\partial q_i}\right)$$

$$= 2\sum_m m\frac{\partial(x,y)}{\partial(q_j,q_i)}\dot{q}_j \qquad \text{summed also over } j. \qquad (18)$$

The factors $\dfrac{\partial(x,y)}{\partial(q_j,q_i)}$ are real geometrical quantities associated with the system and can be regarded as of known analytical form when the generalized coordinates have been selected. We may denote them by β_{ij}, so that we have

$$\beta_{ij} = 2\sum_m m\left(\frac{\partial x}{\partial q_i}\cdot\frac{\partial y}{\partial q_j} - \frac{\partial y}{\partial q_i}\cdot\frac{\partial x}{\partial q_j}\right) = -\beta_{ji} \qquad (19)$$

and $\beta_{ii} = 0$, for each value of i, no summation being involved.

Substituting the value of T given by (12) in the left-hand side of equations (15), we have

$$\frac{d}{dt}\left(\frac{\partial T_R}{\partial \dot{q}_i}\right) - \frac{\partial T_R}{\partial q_i} + \omega\left\{\frac{d}{dt}\left(\frac{\partial H_R}{\partial \dot{q}_i}\right) - \frac{\partial H_R}{\partial q_i}\right\} - \tfrac{1}{2}\omega^2\frac{\partial I}{\partial q_i} = -\frac{\partial V}{\partial q_i} \quad (i = 1, 2, ..., n). \qquad (20)$$

The term on the left multiplying ω has just been calculated, and inserting this in terms of the β_{ij}'s, the equations become

$$\frac{d}{dt}\left(\frac{\partial T_R}{\partial \dot{q}_i}\right) - \frac{\partial T_R}{\partial q_i} + \omega(\beta_{1i}\dot{q}_1 + \beta_{2i}\dot{q}_2 + ... + \beta_{ni}\dot{q}_n) = -\frac{\partial}{\partial q_i}(V - \tfrac{1}{2}\omega^2 I) + (F_i). \qquad (21)$$

These are the equations of motion of the system relative to uniformly rotating axes. They differ in two important respects from the usual equations referred to unaccelerated axes. First, by the presence of the term $-\tfrac{1}{2}\omega^2 I$, which appears added to the potential energy. The force terms resulting from this are sometimes called the *ordinary centrifugal forces*. Second, by the terms in the velocities, $\omega\beta_{ij}\dot{q}_i$, which depend also on the rotation ω; these are usually called *gyroscopic terms*, but sometimes also the *compounded centrifugal forces*. Their presence is of special importance, for they can, if sufficiently strong, render a system stable in a certain sense (to be explained later) when the usual energy conditions would indicate instability.

The terms F_i have been added in (21) to represent any external forces that might be acting. It will be convenient later in discussing the possible effects of friction to identify these terms with the frictional forces.

Conditions for relative equilibrium

If a configuration is one of equilibrium relative to the uniformly rotating frame, then we shall have $T_R = 0$ and all the velocities $\dot{q}_i = 0$. Hence in the absence of any external forces the conditions are simply

$$\frac{\partial}{\partial q_i}(V - \tfrac{1}{2}\omega^2 I) = 0 \quad (i = 1, 2, ..., n). \tag{22}$$

Thus the only difference in the conditions for equilibrium, as compared with non-rotating statical systems, is the appearance of the term $-\tfrac{1}{2}\omega^2 I$ added to the potential energy. That is, to find the equilibrium position we simply use $V - \tfrac{1}{2}\omega^2 I$ instead of V and express the condition that it shall be stationary. We may term this function *the total mechanical potential*.

Conditions for stability

If we multiply equations (21) by $\dot{q}_i$ and sum for all i we obtain

$$\frac{d}{dt}\left(\frac{\partial T_R}{\partial \dot{q}_i}\right)\dot{q}_i - \frac{\partial T_R}{\partial q_i}\dot{q}_i + \omega\beta_{ij}\dot{q}_i\dot{q}_j = -\frac{\partial}{\partial q_i}(V - \tfrac{1}{2}\omega^2 I)\dot{q}_i + F_i\dot{q}_i. \tag{23}$$

Since $\beta_{ij} = -\beta_{ji}$, the terms in ω vanish identically. To reduce this equation further, we have, by Euler's theorem, since T_R is a homogeneous quadratic function of the $\dot{q}_i$'s,

$$2T_R = \frac{\partial T_R}{\partial \dot{q}_i}\dot{q}_i$$

and differentiation of this with respect to the time gives

$$2\frac{dT_R}{dt} = \frac{\partial T_R}{\partial \dot{q}_i}\ddot{q}_i + \frac{d}{dt}\left(\frac{\partial T_R}{\partial \dot{q}_i}\right)\dot{q}_i.$$

Also by direct total differentiation of T_R we have

$$\frac{dT_R}{dt} = \frac{\partial T_R}{\partial \dot{q}_i}\ddot{q}_i + \frac{\partial T_R}{\partial q_i}\dot{q}_i.$$

Whence

$$\frac{dT_R}{dt} = \frac{d}{dt}\left(\frac{\partial T_R}{\partial \dot{q}_i}\right)\dot{q}_i - \frac{\partial T_R}{\partial q_i}\dot{q}_i,$$

so that (23) becomes

$$\frac{d}{dt}(T_R + V - \tfrac{1}{2}\omega^2 I) = F_i\dot{q}_i. \tag{24}$$

If there are no external forces, other than the couple maintaining the constant rotation, which is not of course included in the F_i's, this equation possesses the integral

$$T_R + V - \tfrac{1}{2}\omega^2 I = \text{constant}. \tag{25}$$

Secular stability

For equilibrium we have seen that $V - \frac{1}{2}\omega^2 I$ must be stationary, and this value may always be taken as zero. If, further, the configuration is such that $V - \frac{1}{2}\omega^2 I$ is an absolute minimum, the position is said to be *secularly stable*.

Let us consider now small motion in the neighbourhood of such a position and suppose that the coordinates are chosen, as is always possible, to vanish for the equilibrium form. They can also be transformed to make both T_R and $V - \frac{1}{2}\omega^2 I$ sums of square terms only. This is always possible by means of a real linear non-singular transformation since T_R is necessarily positive definite. We may therefore write, using q_i now to denote the new coordinates,

$$2T_R = a_i \dot{q}_i^2, \qquad \text{where } a_i > 0, \text{ necessarily,}$$

and
$$2(V - \tfrac{1}{2}\omega^2 I) = b_i q_i^2, \qquad \text{where } b_i > 0 \text{ in this case.} \tag{26}$$

If the system is given a slight displacement, the integral (25) will become

$$T_R + \tfrac{1}{2}\Sigma b_i q_i^2 = c, \tag{27}$$

where c is a small positive constant. Since T_R cannot become negative, this means that each coordinate can never exceed a certain amount. For instance, q_1 can certainly never exceed $(2c/b_1)^{\frac{1}{2}}$ and may not reach this value. Thus, in the absence of friction, the system must oscillate in the immediate neighbourhood of the equilibrium position, the amplitudes of the various terms expressing the oscillation all being small. The system will therefore be stable in the ordinary sense of the word used for statical systems. Such (rotating) systems are in fact said to be *ordinarily stable*.

If friction proportional to the relative velocities is present, as it must always be to some extent in natural systems, these oscillations will die away. For the frictional forces may be regarded as external forces acting in such a way that the terms $F_1 \dot{q}_1, F_2 \dot{q}_2, ..., F_n \dot{q}_n$ are necessarily all negative, since friction always operates to decrease the kinetic energy. Equation (24) then implies that

$$\frac{d}{dt}(T_R + V - \tfrac{1}{2}\omega^2 I) = \text{a negative quantity.}$$

Hence, however slight the friction, $T_R + V - \frac{1}{2}\omega^2 I$ must continually decrease so long as any of the velocities $\dot{q}_i$ are different from zero. Thus both T_R and $V - \frac{1}{2}\omega^2 I$ must become zero, which means that all the coordinates fall to zero, and the system returns to the equilibrium position. It is not sufficient for T_R alone to fall to zero, since until $V - \frac{1}{2}\omega^2 I$ also becomes zero the system cannot be permanently in equilibrium and the velocities would not remain zero. Such a system, in which all the coefficients of stability are positive, is said to be *secularly stable*. We therefore see that if an equilibrium configuration of a rotating system is secularly stable it is also necessarily ordinarily stable. Such systems are sometimes said to be 'thoroughly stable'.

The difference between the two kinds of stability will become more clear from the subsequent discussion of unstable systems, for which the distinction is of great importance.

Ordinary stability of rotating systems: small motion

Let us consider next the equations of small motion in the neighbourhood of an equilibrium configuration of the rotating system. Using the quadratic forms (26) for T_R and $V - \frac{1}{2}\omega^2 I$, the equations of motion (21) become, for small values of the q_i's

$$\left.\begin{aligned}
a_1\ddot{q}_1 + b_1 q_1 + \omega(\quad\quad \beta_{12}\dot{q}_2 + \beta_{13}\dot{q}_3 + \ldots) &= F_1, \\
a_2\ddot{q}_2 + b_2 q_2 + \omega(\beta_{21}\dot{q}_1 + \quad\quad \beta_{23}\dot{q}_3 + \ldots) &= F_2, \\
\ldots\quad\quad \ldots\quad\quad \ldots\quad\quad \ldots\quad\quad &= \ldots, \\
a_n\ddot{q}_n + b_n q_n + \omega(\beta_{n1}\dot{q}_1 + \beta_{n2}\dot{q}_2 + \quad \ldots\quad) &= F_n.
\end{aligned}\right\} \quad (28)$$

For $\omega = 0$ these equations reduce to the usual form for principal or normal coordinates. The free oscillations, $F_1 = 0$, of any one of the coordinates is then independent of that of any of the others. But for rotating systems this no longer holds even though both T_R and $V - \frac{1}{2}\omega^2 I$ are reduced to the sums of square terms only. For if any one coordinate, say q_1, varies, the presence of the gyroscopic terms in general brings about consequent changes in all the remaining coordinates.

Investigation of the free oscillations, or small motion, can however be made in the usual way; that is, put $F_i = 0$ and assume a solution of (28) of the form

$$q_i = q_{i0}e^{\lambda t} \quad (i = 1, 2, \ldots, n), \quad (29)$$

where the q_{i0}'s are small constants, to be determined, and the exponential term is the same for every coordinate. This will be a solution provided the following system of linear equations is satisfied:

$$\left.\begin{aligned}
(a_1\lambda^2 + b_1)q_{10} + \quad \omega\lambda\beta_{12}q_{20} + \omega\lambda\beta_{13}q_{30} + \quad\quad \ldots \quad &= 0, \\
\omega\lambda\beta_{21}q_{10} + (a_2\lambda^2 + b_2)q_{20} + \omega\lambda\beta_{23}q_{30} + \quad\quad \ldots \quad &= 0, \\
\ldots\quad\quad \ldots\quad\quad \ldots\quad\quad \ldots\quad &= 0, \\
\omega\lambda\beta_{n1}q_{10} + \quad\quad \ldots \quad\quad\quad \ldots \quad + (a_n\lambda^2 + b_n)q_{n0} &= 0.
\end{aligned}\right\} \quad (30)$$

The expression of the consistency of these linear homogeneous equations in the q_{i0}'s gives rise to a determinantal equation for the possible values of λ, namely

$$\begin{vmatrix}
a_1\lambda^2 + b_1 & \omega\beta_{12}\lambda & \omega\beta_{13}\lambda & \ldots \\
\omega\beta_{21}\lambda & a_2\lambda^2 + b_2 & \omega\beta_{23}\lambda & \ldots \\
\omega\beta_{31}\lambda & \omega\beta_{32}\lambda & a_3\lambda^2 + b_3 & \ldots \\
\ldots & \ldots & \ldots & \ldots
\end{vmatrix} = 0. \quad (31)$$

The small motion of the system accordingly depends on the roots of this equation. If λ is one of the roots, and in general the roots will be non-multiple, the n equations (30) are then consistent, so that any $n-1$ of them can be solved for the ratios of the constants $q_{10} : q_{20} : \ldots : q_{n0}$. One at least of these constants must remain arbitrary, but the remainder will be expressible as multiples of it, the multiples being different for the different values of λ. The complete solution of (28), with $F_i = 0$, will then be given by the linear sum of the $2n$ solutions obtained in this way, one for each of the $2n$ roots of the equation (31) in λ.

Stability

If all the roots in λ are not pure imaginary, terms increasing exponentially with the time will occur for a general small displacement of the system. To show this, we notice first that (31) is in effect an equation in λ^2. For since $\beta_{ij} = -\beta_{ji}$, if $-\lambda$ is written for λ, the rows and columns are interchanged. Had ω not been present, only the terms in the leading diagonal would have occurred, and the roots would be simply

$$\lambda^2 = -b_1/a_1, \quad -b_2/a_2, \quad ..., \quad -b_n/a_n, \tag{32}$$

so that stability would require $b_1, b_2, ..., b_n$ to be all positive. But now the determinantal equation no longer has simple roots like this, and in general roots in λ^2 other than purely real ones will occur in pairs of the form

$$\lambda^2 = \gamma \pm i\delta \qquad (i = \sqrt{-1}),$$

and these will give individual values of λ of the form

$$\lambda = \pm (p \pm iq).$$

The time factors in the coordinates, when put in real form, with these values of λ will be of the form $$A\,e^{pt} \cos(qt + \alpha) + B\,e^{-pt} \cos(qt + \beta),$$

and a general oscillation, as opposed to one specially started, will contain all these terms. Accordingly, if $p \neq 0$, one or other of the amplitudes $A\,e^{pt}$ or $B\,e^{-pt}$ will increase indefinitely. The motion will not remain small and the system will be *ordinarily unstable*.

The condition for the system to be ordinarily stable is therefore that, regarded as an equation in λ^2, the determinantal equation (31) must have all its roots real and negative.

Stated in this form, the condition is the same as for non-rotating systems, as expressed by (32), but it is particularly to be noticed that for rotating systems this no longer requires all the coefficients of stability $b_1, b_2, ..., b_n$ to be positive. For if $-\lambda_1^2, -\lambda_2^2, ..., -\lambda_n^2$ are the (supposed) real and negative roots of the equation in λ^2, then by considering the term of highest degree and the constant term in (31) we get

$$b_1 b_2 ... b_n = a_1 a_2 ... a_n . \lambda_1^2 \lambda_2^2 ... \lambda_n^2. \tag{33}$$

Hence, since $a_1, a_2, ..., a_n$ are all essentially positive, the sign of the product $\lambda_1^2 \lambda_2^2 ... \lambda_n^2$, which must be positive for stability, will be the same as that of the product $b_1 b_2 ... b_n$. Thus, as far as this requirement is concerned, ordinary stability may hold even though some of the b_i's are negative, but the number of negative coefficients of stability must nevertheless be even. We thus have the following important result (for systems involving a finite number of degrees of freedom):

If a finite system becomes secularly unstable as a result of the changing of sign of one (or an odd number) of the coefficients of stability, it must simultaneously become ordinarily unstable.

The foregoing discussion does not prove that a system can be ordinarily stable when $V - \frac{1}{2}\omega^2 I$ is a maximum with regard to some of the coordinates, but only that negative coefficients of stability, if even in number, are not inconsistent with ordinary stability. To establish the result the roots of (31) would have to be more fully investigated, and this would require detailed knowledge or assumption as to the values of the β_{ij}'s and ω. But it is not necessary for our purpose to consider the question with full generality; instead we show below, for the case of two degrees of freedom, that a system can be ordinarily stable when $V - \frac{1}{2}\omega^2 I$ is a maximum.

Systems of an infinite number of degrees of freedom

The present discussion concerns systems of a finite number of degrees of freedom, whereas for a liquid the number of freedoms is infinite. It has been shown by Hilbert that the theory can, however, be rigorously extended to infinite systems, and the occurrence of imaginary or complex solutions, so that terms of the form $e^{\pm pt}$ appear in the coordinates, can again be regarded as indicating instability as for finite systems. The equation for λ will in general have an infinite number of roots and only in special cases will it be representable in algebraic form, but it so happens that in the problem with which we shall be concerned the period equation when suitably derived breaks up into an infinite series of equations each of algebraic form.

It is important to remember, however, that results derived above, such as that concerning the number of coefficients of stability that change sign and its relation to stability, will not necessarily carry over unchanged for infinite systems. To avoid any possibility of error arising in such a way, the safer plan is to examine each particular problem on its own rather than to infer its properties from results established only for finite systems.

Condition for loss or gain of ordinary stability

Suppose we have a configuration of a rotating system that is secularly stable, and therefore also ordinarily stable, let us consider what may happen as the system evolves along a linear series. In general all the roots $-\lambda_1^2, -\lambda_2^2, \ldots, -\lambda_n^2$ of (31) will be different, so that if we regard them as arranged in order of magnitude, that is, $\lambda_1^2 < \lambda_2^2 < \ldots < \lambda_n^2$, the first root that may change sign will be $-\lambda_1^2$. But this is equivalent to the product $b_1 b_2 \ldots b_n$ changing sign. Hence ordinary stability is certainly lost when one, or an odd number, of the coefficients of stability changes sign, whereas it *may* continue to hold if an even number of coefficients of stability change sign simultaneously.

Secular stability cannot hold after ordinary stability is lost, but it may hold right up to this point and usually will, since in general only one coefficient of stability will change sign at a time. To sum up, the relation between secular stability and ordinary stability may be stated in the following four propositions:

(i) *If an equilibrium configuration is secularly stable it is also ordinarily stable.*

(ii) *If it is secularly unstable, it may be ordinarily stable or it may be ordinarily unstable.*

(iii) *If it is ordinarily stable, it may be secularly stable or unstable.*

(iv) *If it is ordinarily unstable, it is necessarily secularly unstable.*

These are not, of course, four independent propositions; for example, (iv) obviously follows from (i).

It has seemed worthwhile to go into the relation between the two kinds of stability in some detail because it is over this very question that former writers appear to have fallen into serious error. For instance, Jeans says:*

> 'Further, we see that as the physical conditions of a system gradually change, secular instability necessarily sets in *before* ordinary stability. Thus for problems of cosmogony it is secular instability alone which is of interest. A system never attains to a configuration in which ordinary instability comes into operation, since secular instability must always have previously intervened.'

The present discussion makes it perfectly clear that these statements are incorrect. Jeans's subsequent discussion of the consequences of his solution of the problem rest on the same misapprehension as to what is involved.

The nature of secular instability

If we have a configuration that is ordinarily stable but secularly unstable, let us consider how it will develop. In the complete absence of friction it will, if slightly disturbed, simply oscillate about the equilibrium position. But if friction is present, we shall have by (24)

$$\frac{d}{dt}(T_R + V - \tfrac{1}{2}\omega^2 I) = -\text{ve quantity.}$$

In equilibrium we have $T_R = 0$, and also $V - \tfrac{1}{2}\omega^2 I$ is a maximum (with respect to some of the coordinates) which may conveniently be taken as zero. We can therefore imagine the system displaced to an adjacent configuration in which $V - \tfrac{1}{2}\omega^2 I$ is negative, so that initially

$$T_R + V - \tfrac{1}{2}\omega^2 I = -c,$$

where c is a small positive constant. In the subsequent motion either T_R, $V - \tfrac{1}{2}\omega^2 I$, or both these quantities will be different from zero, and hence in any event the system will move. But when motion occurs the presence of friction causes the left-hand side to decrease further, and after a small interval t we shall have
$$T_R + V - \tfrac{1}{2}\omega^2 I = -c - kt, \qquad \text{where } k > 0.$$

Thus even if T_R should vanish subsequently, $V - \tfrac{1}{2}\omega^2 I$ would then be still further from the equilibrium value, further motion would therefore necessarily develop, and the left-hand side would continue to decrease. Since T_R cannot be negative, this means that $V - \tfrac{1}{2}\omega^2 I$ becomes more and more negative, and the system therefore departs more and more from the original equilibrium configuration.

It is important to notice that the rate of departure clearly depends on the amount of friction present and vanishes with it. Hence the departure from equilibrium cannot be assumed to be necessarily a rapid one, though however small the friction the amplitude of the motion must increase indefinitely given time enough.

* *Astronomy and Cosmogony*, p. 199.

The nature of ordinary instability

Ordinary instability means that the expressions for the initial motion of the system involve exponential terms of the form $e^{\pm pt}$ where p is real and depends only on the geometry of the system, as it were, and will in general be a quantity of standard magnitude. The rate of departure from the equilibrium position will therefore in general become rapid, and the system will continue to move until, if friction is present, a new configuration of relative equilibrium is reached. But the equations of small motion indicating the instability can reveal only the initial stages of this motion and will not in general give any indication of the new steady state to which the system will eventually move.

Systems of two degrees of freedom

The following investigation, due to Lamb, shows how the foregoing theory applies for systems of two degrees of freedom, and establishes that a system can be ordinarily stable while secularly unstable.

Neglecting friction to begin with and absorbing ω into the β's and the positive quantities a_1, a_2 into the coordinates, the equations of small motion may be put in the form

$$\ddot{q}_1 - \beta \dot{q}_2 + b_1 q_1 = 0,$$
$$\ddot{q}_2 + \beta \dot{q}_1 + b_2 q_2 = 0.$$

Assuming in the usual way $q_1 = q_{10} e^{\lambda t}$, $q_2 = q_{20} e^{\lambda t}$, and eliminating the ratio $q_{10} : q_{20}$, we find

$$\lambda^4 + (b_1 + b_2 + \beta^2) \lambda^2 + b_1 b_2 = 0. \tag{34}$$

The two values of λ^2 will be real if

$$(b_1 + b_2 + \beta^2)^2 > 4 b_1 b_2,$$

or

$$\beta^2 (\beta^2 + 2b_1 + 2b_2) + (b_1 - b_2)^2 > 0. \tag{35}$$

There are three cases to consider:

(i) b_1 *and* b_2 *both positive.* Condition (35) is certainly fulfilled since both terms are positive for any value of β. Also both roots of (34) are negative since

$$\lambda_1^2 \lambda_2^2 = b_1 b_2 > 0,$$
$$\lambda_1^2 + \lambda_2^2 = -(b_1 + b_2 + \beta^2) < 0.$$

The system is therefore ordinarily stable whatever the value of β.

(ii) b_1 *and* b_2 *of opposite sign.* The first form of (35) shows, in the present case, that the requirement is necessarily satisfied. But the roots of (34) must be of opposite signs since $\lambda_1^2 \lambda_2^2$ is now negative. The positive root then leads to terms of the form $e^{\pm pt}$, and hence it follows that the system is ordinarily unstable.

(iii) b_1 *and* b_2 *both negative.* In this case (35) will obviously hold if β^2 is large enough. If we suppose $2b_1 + 2b_2 + \beta^2 > 0$, then

$$\lambda_1^2 \lambda_2^2 > 0 \quad \text{and} \quad \lambda_1^2 + \lambda_2^2 < 0.$$

Both roots are therefore essentially negative and the system ordinarily stable. The modified potential energy function is however now a maximum.

These results are in accord with the earlier investigation, and the last case demonstrates the possibility of ordinary stability holding in a secularly unstable system.

Effect of friction

Let us consider how the motion will be modified in case (iii) if frictional forces proportional to the velocities are present. It will be shown that in a certain sense the system is now unstable.

The equations of motion may be written

$$\ddot{q}_1 + b_1 q_1 + f_{11}\dot{q}_1 + (f_{12} - \beta)\dot{q}_2 = 0,$$

$$\ddot{q}_2 + b_2 q_2 + (f_{12} + \beta)\dot{q}_1 + f_{22}\dot{q}_2 = 0,$$

where f_{11}, f_{12}, f_{22} are small quantities, the coefficients of the so-called 'dissipation function' given by $\frac{1}{2}(f_{11}\dot{q}_1^2 + 2f_{12}\dot{q}_1\dot{q}_2 + f_{22}\dot{q}_2^2)$. This function is always positive, so that work is always being done against the system, reducing its energy. Accordingly

$$f_{11} > 0, \quad f_{22} > 0, \quad \text{and} \quad f_{11}f_{22} > f_{12}^2.$$

The equation for the periods now becomes

$$\lambda^4 + (f_{11} + f_{22})\lambda^3 + (b_1 + b_2 + \beta^2 + f_{11}f_{22} - f_{12}^2)\lambda^2 + (f_{11}b_2 + f_{22}b_1)\lambda + b_1 b_2 = 0, \quad (36)$$

which differs from (34) by the presence of the various terms in the coefficients of the dissipation function.

The original roots being $\pm i\lambda_1, \pm i\lambda_2$ we may suppose the modified roots to be

$$p_1 \pm i(\lambda_1 + r_1), \quad p_2 \pm i(\lambda_2 + r_2),$$

where p_1, p_2, r_1, r_2 must all be small, since they vanish with the f_{ij}'s. The sum of the four roots gives at once

$$2p_1 + 2p_2 = -f_{11} - f_{22} < 0,$$

and the sum of the reciprocals of the roots gives, to the first order in the small quantities,

$$\frac{2p_1}{\lambda_1^2} + \frac{2p_2}{\lambda_2^2} = -\frac{f_{11}}{b_1} - \frac{f_{22}}{b_2} > 0,$$

since b_1 and b_2 are both negative. Hence p_1 and p_2 are of opposite signs, and therefore if the system is disturbed one of the oscillations will die away, and the other will increase exponentially at a rate depending on the f_{ij}'s.

Eliminating p_1, say, we find

$$p_2(\lambda_1^2 - \lambda_2^2) > 0.$$

Hence if $\lambda_1 < \lambda_2$, p_2 is negative and p_1 is positive. Accordingly that oscillation increases (p_1 positive) corresponding to the smaller frequency (here λ_1), that is, the one corresponding to the longer of the two original periods. The amplitude of the oscillation of shorter period decays.

Stability when the angular momentum remains constant

So far we have been considering dynamical systems in which ω is maintained constant and oscillations can be regarded as taking place relative to these rotating axes. If, however, we have a freely rotating system in relative equilibrium, so that ω is constant for this state, it is conceivable that if given a small displacement, the subsequent average rate of rotation would be slightly different from ω, but that in a frame rotating at this different rate, the motion is nevertheless expressible by real harmonic terms. The position is that whereas an ignorable coordinate, such as an orbital angle, can increase indefinitely without necessarily producing great change in the general form of the path, any palpable coordinate must change only slightly under disturbance for a system to be stable.

For instance, a particle describing a circular orbit under the inverse square law will be at relative rest in a certain rotating frame. If it is given a slight tangential impulse, its total energy will change and therefore also its mean motion round the centre of force. In the original frame it will therefore depart further and further from its equilibrium position and an investigation of its stability based on $V - \frac{1}{2}\omega^2 I$ would suggest instability.

On the other hand, the condition

$$\delta(V - \tfrac{1}{2}\omega^2 I) > 0$$

with ω assumed constant is always a *sufficient* condition for stability, but not a necessary one. Accordingly, if stability is indicated by means of this function, then stability is assured, but if instability is indicated it may not mean a physically real instability, and further examination of the problem may be necessary. A more general criterion will next be obtained.

Freely rotating systems

Let us refer the system to a set of rectangular axes Ox, Oy, Oz rotating about Oz with angular velocity ω, as yet unspecified and not assumed constant. Then we have seen that

$$T = T_R + \omega H_R + \tfrac{1}{2}\omega^2 I,$$
$$H = H_R + \omega I.$$

If the z-axis is taken as the invariable angular momentum axis, then H is constant for all oscillations or motions of the system. Suppose now that we choose ω always so that the relative angular momentum H_R is zero. This defines a certain ω, which will not necessarily or usually be constant with time, but which reduces to the usual rigid-body angular velocity when the system is in a steady state. We then have

$$T = T_R + \tfrac{1}{2}\omega^2 I, \tag{37}$$
$$H = \omega I, \tag{38}$$

and elimination of the variable ω gives

$$T = T_R + H^2/2I. \tag{39}$$

For a freely rotating system, if V is the potential of the external forces, the energy equation $T + V = $ constant, now assumes the form

$$T_R + V + H^2/2I = \text{constant}, \tag{40}$$

which is similar in form to (25) except that the constant H now appears in place of ω, and the term $H^2/2I$ replaces the former term $-\frac{1}{2}\omega^2 I$.

Exactly similar arguments to those already given for systems with constant rotation now show that systems of constant angular momentum will be secularly stable if $V + H^2/2I$ is an absolute minimum, and therefore also ordinarily stable. Otherwise, the system will be secularly unstable and may or may not be ordinarily stable.

The importance of this criterion was first pointed out by Schwarzschild.

Cases in which the two criteria are equivalent

It can readily be shown that if the small displacements of the system from the equilibrium configuration are such that the moment of inertia I is unchanged to the first order of small quantities (that is, changed only in second order) the two criteria are equivalent.

For since H is invariable, we have $\delta(\omega I) = 0$, and hence

$$\delta\omega = -\frac{\omega}{I}\,\delta I,$$

and is also of second order of smallness. We then have that

$$\delta\left(V + \frac{H^2}{2I}\right) = \delta V - \frac{H^2}{2I^2}\,\delta I$$

$$= \delta V - \tfrac{1}{2}\omega^2\,\delta I,$$

correct to the second order of small quantities. Hence the condition for stability may be obtained by consideration of $\delta V - \tfrac{1}{2}\omega^2\,\delta I$, which is exactly as if the function $V - \tfrac{1}{2}\omega^2 I$ were used with ω treated as constant. Thus the conditions of secular stability are equivalent in these circumstances.

Where ordinary stability is concerned, since ω is unaltered except for second-order changes, the terms in the first-order oscillation equations, corresponding to (28), that involve $\delta\omega$ can be neglected, since they are of second order. Thus constant ω may again be assumed in considering small oscillations, so that the conditions for ordinary stability may also be derived on this basis.

EXAMPLES OF STABILITY OF ROTATING SYSTEMS

(i) *A particle of unit mass moves under a central attractive force which at distance r is of amount μ/r^n, where n is regarded as a slowly changing parameter.*

Here $V = -\mu/(n-1)r^{n-1}$, and $I = r^2$. Hence if h denotes the orbital angular momentum, stability will depend on the function

$$U = V + \frac{h^2}{2I} = -\frac{\mu}{(n-1)r^{n-1}} + \frac{h^2}{2r^2}.$$

Configurations of relative equilibrium are therefore given by

$$\frac{dU}{dr} = \frac{\mu}{r^n} - \frac{h^2}{r^3} = 0.$$

This gives $h^2 = \mu/r^{n-3}$, which is simply the condition for a circular orbit of radius r.

To examine its secular stability we have

$$\frac{d^2U}{dr^2} = -\frac{n\mu}{r^{n+1}} + \frac{3h^2}{r^4} = \frac{h^2}{r^4}(3-n).$$

Hence such an orbit will be secularly stable if $n < 3$.

The conditions implied by this calculation would correspond physically to the presence of a resisting medium rotating round the centre of force in such a way that the particle had zero velocity relative to it only when moving in a circular orbit. If $n < 3$, a slightly non-circular orbit would have its radial motion gradually damped out, and the system would be secularly stable.

It may easily be shown that this is also the condition for ordinary stability. For if in the radial equation of motion

$$\ddot{r} - r\dot{\theta}^2 = -\mu/r^n$$

we put $r = a + \xi$, where ξ is small and a is the radius of the undisturbed orbit, and eliminate the angular rate $\dot{\theta}$ by means of $r^2\dot{\theta} = h + \delta h$, say, this gives

$$\ddot{\xi} + \left(\frac{3h^2}{a^4} - \frac{\mu n}{a^{n+1}}\right)\xi = \frac{\delta h}{a^3},$$

and the condition that the coefficient of ξ is positive reduces to $n < 3$.

It is to be noticed that because of the presence of the term $\delta h/a^3$ on the right-hand side, the oscillations are not about $r = a$ as mean value but about a slightly different value (if $\delta h \neq 0$), and the mean motion in the disturbed path will therefore differ to the first order from the original value. Thus if measured in a frame rotating with the original angular velocity, the motion would appear to be unstable, though in fact it would not represent a true instability in the sense that a palpable coordinate increasing indefinitely would do.

(ii) *A particle of unit mass moves under gravity on the inner surface of a spherical bowl of radius a that is made to rotate round its vertical axis with constant angular velocity ω.*

If θ denotes the inclination to the vertical of the radius drawn to the particle, the function appropriate to the system will be

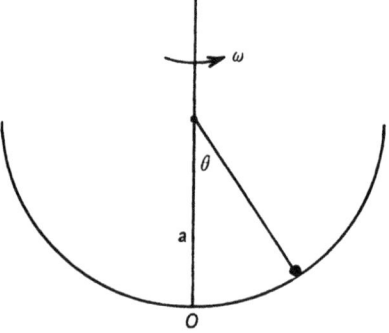

$$U = V - \tfrac{1}{2}\omega^2 I = -ga\cos\theta - \tfrac{1}{2}\omega^2 a^2 \sin^2\theta.$$

The equilibrium positions are therefore given by

$$\frac{dU}{d\theta} = ga\sin\theta - \omega^2 a^2 \sin\theta\cos\theta = 0.$$

Fig. 7.

Hence $\theta = 0$, or $\cos\theta = g/a\omega^2$, a position that exists only if $a\omega^2 > g$.

The position $\theta = 0$ is evidently ordinarily stable always, since if friction is ignored the rotation of the bowl has no influence. In the position given by $\cos\theta = g/a\omega^2$, the particle revolves as a conical pendulum and no friction is required to maintain it in this position. Also such a motion is well known to be ordinarily stable.

To consider the secular stability of these positions, we have

$$\frac{d^2 U}{d\theta^2} = ga\cos\theta - \omega^2 a^2(\cos^2\theta - \sin^2\theta).$$

For $\theta = 0$, this gives $U_{\theta\theta} = ga - \omega^2 a^2 > 0$, and hence the lowest position is stable so long as $a\omega^2 < g$. But when the upper position becomes possible, the lowest position is secularly unstable. Also, for $\cos\theta = g/a\omega^2$, $U_{\theta\theta} = \omega^2 a^2 \sin^2\theta > 0$, and hence the upper position is always secularly stable. At $\omega^2 = g/a$, there occurs a transfer of stability to a new series of configurations, and the system therefore provides an example of case (i) of the theory of exchange of stabilities (p. 8).

Initial motion of the particle near the lowest position

A discussion of the small motion in the neighbourhood of $\theta = 0$, when this position is secularly unstable, affords an interesting example of the results reached on pp. 24–25.

The equations of small motion relative to horizontal rectangular axes Ox, Oy, rotating with the bowl may be written

$$\ddot{x} - 2\omega\dot{y} - \omega^2 x = -2k\dot{x} - n^2 x,$$

$$\ddot{y} + 2\omega\dot{x} - \omega^2 y = -2k\dot{y} - n^2 y,$$

where k is a small positive constant representing the friction, and $n^2 = g/a$. Writing $\xi = x + iy$, these equations can be combined into

$$\ddot{\xi} + (2i\omega + 2k)\dot{\xi} + (n^2 - \omega^2)\xi = 0.$$

Assuming $\xi \propto e^{\lambda t}$, the period equation is

$$\lambda^2 + (2i\omega + 2k)\lambda + (n^2 - \omega^2) = 0,$$

or
$$(\lambda + i\omega)^2 = -n^2 - 2k\lambda.$$

Hence to the first order in k we have

$$\lambda = -i\omega \pm in - k\left(1 \mp \frac{\omega}{n}\right).$$

Hence
$$x + iy = e^{-i\omega t}\left\{A\, e^{int}.e^{-k(1-\frac{\omega}{n})t} + B\,e^{-int}.e^{-k(1+\frac{\omega}{n})t}\right\}.$$

Relative to fixed axes at the lowest point the coordinates are simply the real and imaginary parts of $(x + iy)e^{i\omega t}$. Clearly, if $\omega < n$, that is if $a\omega^2 < g$, both real exponential terms decay, and the lowest position is stable, as already shown. But if $\omega > n$, that is $a\omega^2 > g$, the first exponential term gradually increases, so that the distance from the origin must increase. The term of increasing amplitude corresponds to the oscillation of longer period, $2\pi/(n - \omega)$, in the motion that would obtain if k were zero. The amplitude of the term of period $2\pi/(n + \omega)$ always decays.

(iii) *A particle moves on a wire in the form of a curve possessing an asymptote at* $x = a$, *and is attracted towards the z-axis with a force of amount* λx^3.

Here

$$U = V - \tfrac{1}{2}\omega^2 I = \tfrac{1}{4}\lambda x^4 - \tfrac{1}{2}\omega^2 x^2.$$

Equilibrium positions are therefore given by

$$\frac{dU}{dx} = \lambda x^3 - \omega^2 x = 0,$$

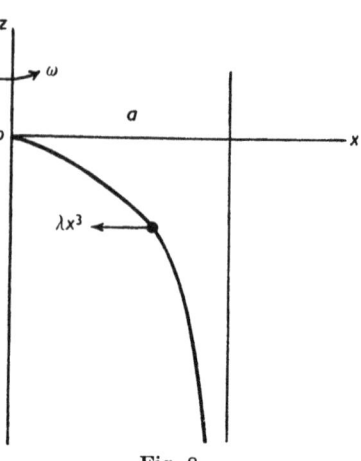

Fig. 8.

that is, $x = 0$, or $x = \sqrt{(\omega^2/\lambda)}$. Hence there is a real position of equilibrium, other than $x = 0$, only if $\omega^2 < \lambda a^2$.

To examine the stability of this position, we have

$$\frac{d^2 U}{dx^2} = 3\lambda x^2 - \omega^2.$$

Hence the position at $x = 0$ is unstable if $\omega > 0$, and the position at $x = \sqrt{(\omega^2/\lambda)}$ is stable. Thus if ω slowly increases there is a linear series of equilibrium configurations that terminates at $x = a$, where $\omega^2 = \lambda a^2$.

Where ordinary stability is concerned it is readily seen that the position $x = 0$ is ordinarily unstable, and the other position ordinarily stable, and this is what would be expected for a system of one degree of freedom.

The fundamental assumption of applied mathematics

It may perhaps be worthwhile, in conclusion of this chapter, to mention explicitly the general assumption that underlies almost the whole of applied mathematics. It is the assumption that the accurate solution of approximate (usually linear) differential equations of motion is approximately equal to whatever would be the solution, if this were obtainable, of the accurate (non-linear) equations governing the system. There is certainly no rigorous general mathematical justification of this step, but it has long since represented the standard procedure of applied mathematics, and indeed, quite apart from its seeming naturalness, to certain minds at least, is forced on us by the insuperable difficulties usually lying in the way of any other method of solution. There is, however, the *a posteriori* justification arising from the now numerous problems in which a subsequent comparison with observation or experiment has been made showing considerable measure of agreement. But nevertheless, from the point of view of logical procedure, the step is an assumption without as yet any rigid mathematical demonstration of its validity.

Chapter III

THE SPHERICAL FORM

If a gravitating mass of liquid of uniform density has no angular momentum it is evident intuitively that its equilibrium form will be spherical.

However, to show that the sphere is a possible form, let p denote the pressure, ρ the density, and V the gravitational potential at any point of the liquid. The condition for hydrostatic equilibrium is

$$dp = \rho dV. \tag{1}$$

Since ρ is constant this has the integral

$$p = \rho V + \text{constant}. \tag{2}$$

For a uniform sphere V is constant over the surface, and hence the surface condition, $p = 0$, will be satisfied for an appropriate value of the constant. Moreover, when (2) holds, (1) must also hold at all points of the liquid, and hence the sphere is a possible equilibrium form.

That it is the only form has been proved by Lichtenstein, who shows by simple considerations that any non-rotating equilibrium form must be symmetrical with regard to planes through its mass centre. The sub-stance of the proof is as follows. Suppose the mass divided up into elementary cylindrical columns of infinitesimal cross-sections and with their axes all in the direction of the normal to some plane through O the mass centre. This plane may be taken as Oxy and the direction that of Oz. The mid-points $P_1, P_2, \ldots$ of the columns will define a certain surface. If this surface is plane, the property certainly holds that the configuration is symmetrical about it. Assuming therefore that it is not plane, there must be some point M on it at which the value of z is a maximum, say. (The argument would apply equally well with slight changes if it were a minimum.) If, as in Fig. 9, AB is the column through M, then in relation to every other column the point A is at greater gravitational potential than the point B. Hence for the whole mass $V(A) > V(B)$, and therefore by (2) $p(A) > p(B)$, whereas for an equilibrium configuration the pressures at points on the surface must be equal. Hence there cannot be a point such as M at which z is a maximum, or minimum. The surface must therefore be plane and the form symmetrical about it. For a non-rotating system every plane through O must therefore be a plane of symmetry, and the equilibrium form must be a sphere. Also, since the density is uniform and fixed, this sphere must be unique.

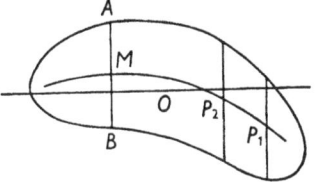

Fig. 9.

It may be mentioned here that the proof remains valid if the mass is rotating about any axis parallel to Oz, for the contribution of the centrifugal force to the potential is the same at A and B. Hence for rotating systems the plane through the mass-centre perpendicular to the axis of rotation must always be a plane of symmetry of an equilibrium form.

Stability

It is also intuitively evident that the spherical form will be thoroughly stable for small displacements, but this can be readily established by means of spherical harmonic analysis, and the proof affords a simple illustration of the general method that is employed to deal with rotating systems.

Suppose a denotes the spherical radius and that after a small displacement at the free surface, without change of total volume, the radius at the point of spherical polar coordinates (r, θ, ϕ) is given by

$$r = a\{1 + s(\theta, \phi)\} \tag{3}$$

where s is everywhere small and is expressible as a sum of surface harmonics, thus

$$s = \sum_{n=1}^{\infty} \sum_{m=1}^{2n+1} A_n^m S_n^m(\theta, \phi), \tag{4}$$

there being $2n+1$ independent functions S_n^m, and where A_n^m are constants small compared with unity. It will be convenient to write this more shortly as

$$s = \Sigma A_n S_n. \tag{5}$$

To the first order of small quantities, the gravitational potential at external points due to a uniform mass whose boundary is the surface (3) is, by a well-known result,

$$ga^2 \left\{ \frac{1}{r} + \sum_{1}^{\infty} \frac{3a^n A_n S_n}{(2n+1)r^{n+1}} \right\}, \tag{6}$$

where g is the value of surface gravity on the original sphere $r = a$. On the surface (3) the value of the potential apart from a constant can be written to the first order

$$2ga \sum_{1}^{\infty} \frac{n-1}{2n+1} A_n S_n. \tag{7}$$

To find the difference $V - V_0$ of the potential energies in the displaced and spherical forms, we adopt the usual method of supposing the free surface to be $r = a(1 + ks)$, where k is a constant, and that a layer of depth $adk.s$ is added to this, the final form being gradually built up by increasing k from 0 to 1. This gives for the work required, correct to the second order,

$$\int_0^1 k\, dk . 2ga \int\!\!\int \left(\sum_1^\infty \frac{n-1}{2n+1} A_n S_n \right) \left(\sum_1^\infty A_n S_n \right) d\sigma,$$

where the double integral is to be evaluated over the surface of the sphere $r = a$. By the orthogonality properties of the surface harmonics this reduces to

$$V = V_0 + ga \sum_1^\infty \frac{n-1}{2n+1} A_n^2 \int\!\!\int S_n{}^2 d\sigma. \tag{8}$$

The coordinates of the system, infinite in number, can be regarded as the A_n's, there being $2n+1$ such coordinates involved in a general displacement of order n. The coefficients of stability are accordingly

$$2ga\,\frac{n-1}{2n+1}\iint S_n^{m2}\,d\sigma, \tag{9}$$

and these are essentially positive for all $n>1$. The potential energy of the sphere is therefore an absolute minimum and the configuration therefore secularly stable.

If $n=1$, the three coefficients of stability vanish, as would be expected, for the harmonics are then x, y and z, and correspond to small translations without change of shape of the sphere as a whole parallel to the coordinate axes. For such displacements the spherical form is evidently neutral.

Small oscillations

If we assume that oscillations are brought about by external gravitational disturbances the resulting motion will be irrotational. If Φ is its velocity potential, the equation of continuity will be Laplace's equation,

$$\nabla^2\Phi = 0,$$

since ρ is uniform. The appropriate general solution will be of the form

$$\Phi = \sum_1^\infty \sum_1^{2n+1} B_n^m \frac{r^n}{a^n} S_n^m(\theta,\phi),$$

terms with negative powers of r being excluded, since these would lead to infinities in the velocity. As before, it will be convenient to write this solution

$$\Phi = \Sigma B_n \frac{r^n}{a^n} S_n.$$

Since the oscillations are considered small, the coefficients B_n will all be small.

At the free surface, we shall have, to the first order,

$$\frac{\partial r}{\partial t} = -\frac{\partial \Phi}{\partial r},$$

and therefore

$$a\dot{A}_n = -\frac{n}{a}B_n. \tag{10}$$

We have already seen that the potential W, say, at the free surface is, apart from a constant, given by (7), so that substituting in the pressure equation, namely

$$\frac{p}{\rho} = \frac{\partial \Phi}{\partial t} - W + \text{constant},$$

and expressing the condition that p is constant at the surface, we get, equating coefficients of the variable functions S_n^m to zero,

$$\dot{B}_n = 2ga\,\frac{n-1}{2n+1}A_n. \tag{11}$$

Hence for the part $A_n^m S_n^m = s_n^m$, say, of the displacement we have

$$\frac{\partial^2 s_n^m}{\partial t^2} + \frac{2n(n-1)}{2n+1} \frac{g}{a} s_n^m = 0. \tag{12}$$

It is evident therefore that for displacements of order n, the $2n+1$ independent surface harmonics S_n^m are normal coordinates for the system. Putting now $s_n \propto e^{i\sigma_n t}$, if the displacement contains only S_n^m, the corresponding frequency σ_n, say, is the same for all S_n^m and is given by

$$\sigma_n^2 = \frac{2n(n-1)}{2n+1} \cdot \frac{g}{a} = \frac{8}{3}\pi G\rho \frac{n(n-1)}{2n+1}, \tag{13}$$

where G is the constant of gravitation.

The frequency is therefore always real and the system accordingly ordinarily stable, as follows in any case from the secular stability already established. It will be noticed that the periods depend only on the density and not on the linear dimensions of the system.

The kinetic energy T of the motion at any stage can be calculated from the formula

$$T = \frac{1}{2}\rho \iint \Phi \frac{\partial \Phi}{\partial \nu} d\sigma$$

taken over the surface, where $\frac{\partial}{\partial \nu}$ denotes normal differentiation. Since Φ is small, this may be written with sufficient accuracy

$$T = \frac{1}{2}\rho \iint \Phi \frac{\partial \Phi}{\partial r} d\sigma$$

$$= \frac{1}{2}\rho a^3 \frac{\dot{A}_n^2}{n} \iint S_n^2 d\sigma. \tag{14}$$

These coefficients are, of course, essentially positive, as we should expect.

The equation of energy for the motion is $T + V = $ constant. But consideration of secular stability requires only the calculation of V, for T must automatically always be positive definite. If V is also essentially positive, none of the co-ordinates can increase beyond a small amount settled by the energy of the initial disturbance.

Chapter IV

THE SPHEROIDAL AND ELLIPSOIDAL FORMS

To demonstrate the existence of the Maclaurin spheroids and Jacobi ellipsoids as possible equilibrium forms we require first an expression for the gravitational potential of such figures at internal points. If we consider an ellipsoid referred to its principal axes as coordinate axes, its equation may be taken as

$$\frac{x^2}{a^2}+\frac{y^2}{b^2}+\frac{z^2}{c^2} = 1, \tag{1}$$

and it will be assumed throughout that $a \geqslant b \geqslant c$. By a well-known result, the gravitational potential at internal points (x, y, z) of a uniform mass bounded by (1) is the quadratic expression in x, y, z

$$W = -\pi G\rho\, abc \int_0^\infty \left(\frac{x^2}{a^2+\lambda}+\frac{y^2}{b^2+\lambda}+\frac{z^2}{c^2+\lambda}-1\right)\frac{d\lambda}{\Delta}, \tag{2}$$

where $\Delta^2 = (a^2+\lambda)(b^2+\lambda)(c^2+\lambda)$, and ρ denotes the density and G the constant of gravitation. This may be written more shortly

$$W = -\pi G\rho(\alpha x^2+\beta y^2+\gamma z^2-\delta), \tag{3}$$

where
$$\alpha = abc \int_0^\infty \frac{d\lambda}{(a^2+\lambda)\Delta}, \quad \beta = abc \int_0^\infty \frac{d\lambda}{(b^2+\lambda)\Delta}, \quad \gamma = abc \int_0^\infty \frac{d\lambda}{(c^2+\lambda)\Delta},$$

and
$$\delta = abc \int_0^\infty \frac{d\lambda}{\Delta}. \tag{4}$$

The potential energy of the whole mass, which will be denoted by V, is given by the volume integral

$$V = -\tfrac{1}{2}\int \rho W d\tau$$

taken throughout the ellipsoid (1). To evaluate this, we have in the first place, by the moment of inertia formulæ for a uniform ellipsoid,

$$\int x^2 d\tau = \tfrac{1}{5}a^2 . \tfrac{4}{3}\pi abc, \quad \int y^2 d\tau = \tfrac{1}{5}b^2 . \tfrac{4}{3}\pi abc, \quad \int z^2 d\tau = \tfrac{1}{5}c^2 . \tfrac{4}{3}\pi abc.$$

Hence
$$V = \tfrac{2}{3}\pi^2 G\rho^2(\tfrac{1}{5}\alpha a^2+\tfrac{1}{5}\beta b^2+\tfrac{1}{5}\gamma c^2-\delta)$$

$$= \tfrac{2}{15}\pi^2 G\rho^2 a^2 b^2 c^2 \int_0^\infty \left(\frac{a^2}{a^2+\lambda}+\frac{b^2}{b^2+\lambda}+\frac{c^2}{c^2+\lambda}-5\right)\frac{d\lambda}{\Delta}.$$

To reduce this, we have

$$2\frac{d\Delta}{\Delta} = \left(\frac{1}{a^2+\lambda}+\frac{1}{b^2+\lambda}+\frac{1}{c^2+\lambda}\right)d\lambda,$$

and therefore

$$\frac{a^2}{a^2+\lambda}+\frac{b^2}{b^2+\lambda}+\frac{c^2}{c^2+\lambda} = 3-\lambda\left(\frac{1}{a^2+\lambda}+\frac{1}{b^2+\lambda}+\frac{1}{c^2+\lambda}\right)$$
$$= 3-2\frac{\lambda}{\Delta}\frac{d\Delta}{d\lambda}.$$

Hence

$$V = -\tfrac{4}{15}\pi^2 G\rho^2 a^2 b^2 c^2 \int_0^\infty \left(1+\frac{\lambda}{\Delta}\frac{d\Delta}{d\lambda}\right)\frac{d\lambda}{\Delta}$$

$$= \tfrac{4}{15}\pi^2 G\rho^2 a^2 b^2 c^2 \int_{\lambda=0}^\infty \left\{\lambda d\left(\frac{1}{\Delta}\right)-\frac{1}{\Delta}d\lambda\right\}$$

$$= \tfrac{4}{15}\pi^2 G\rho^2 a^2 b^2 c^2 \left\{\left[\frac{\lambda}{\Delta}\right]_{\lambda=0}^\infty -2\int_0^\infty \frac{d\lambda}{\Delta}\right\}.$$

But λ/Δ vanishes for both $\lambda = 0$ and $\lambda = \infty$, so that we arrive at the well-known result

$$V = -\tfrac{8}{15}\pi^2 G\rho^2 a^2 b^2 c^2 \int_0^\infty \frac{d\lambda}{\Delta} \tag{5}$$

for the total potential energy of the ellipsoid.

For a spheroid, $a = b$, if we write $c^2 = a^2(1-e^2)$, the foregoing integral can be evaluated in elementary functions, to give

$$V = -\tfrac{16}{15}\pi^2 G\rho^2 (abc)^{5/3} (1-e^2)^{1/6} \frac{\sin^{-1} e}{e}. \tag{6}$$

From the method of calculation, it is seen that the zero of energy in both these results corresponds to infinite dispersion of the matter.

To express these energies in dimensionless form it is convenient to adopt as unit of energy the quantity

$$GM^2/(abc)^{1/3} = (\tfrac{4}{3}\pi\rho)^2 G(abc)^{5/3},$$

and if the unit of length is chosen so that $r = (abc)^{1/3} = 1$, then (5) becomes

$$V = -\tfrac{3}{10}r \int_0^\infty \frac{d\lambda}{\Delta} \qquad \text{for ellipsoids,} \tag{7}$$

and (6) becomes $\qquad V = -\tfrac{3}{5}(1-e^2)^{1/6}\frac{\sin^{-1} e}{e} \qquad$ for spheroids. $\tag{8}$

Condition for a configuration of relative equilibrium

The equations of relative hydrostatic equilibrium of a liquid mass referred to rectangular axes Ox, Oy, Oz rotating round Oz with constant angular velocity ω are

$$\left.\begin{aligned}
\frac{1}{\rho}\frac{\partial p}{\partial x} &= \frac{\partial W}{\partial x}+\omega^2 x,\\
\frac{1}{\rho}\frac{\partial p}{\partial y} &= \frac{\partial W}{\partial y}+\omega^2 y,\\
\frac{1}{\rho}\frac{\partial p}{\partial z} &= \frac{\partial W}{\partial z},
\end{aligned}\right\} \tag{9}$$

where p is the pressure, ρ the density (supposed uniform), and $W(x, y, z)$ is the gravitational potential of the mass in the equilibrium configuration. This system of equations has the integral

$$\frac{p}{\rho} = W + \tfrac{1}{2}\omega^2(x^2 + y^2) + \text{constant}, \tag{10}$$

and if the configuration is one of relative equilibrium it must be possible to choose the constant to make p vanish on the surface of the distribution. When this is done, equation (10) determines p everywhere, and the equilibrium conditions (9) are each automatically satisfied at all points of the liquid. Hence the necessary and sufficient condition that a given configuration, with gravitational potential W, shall be an equilibrium form with angular velocity ω is that the expression

$$W + \tfrac{1}{2}\omega^2(x^2 + y^2) \tag{11}$$

shall be constant over its surface.

If we denote this expression by $\phi(x, y, z)$, say, and the equation of the surface by $f(x, y, z) = 0$, the foregoing condition is simply that $d\phi = 0$ for all displacements such that $f = 0$ and $df = 0$ simultaneously. That is

$$\frac{\partial \phi}{\partial x} dx + \frac{\partial \phi}{\partial y} dy + \frac{\partial \phi}{\partial z} dz = 0$$

for all dx, dy, dz on the surface $f = 0$ such that

$$\frac{\partial f}{\partial x} dx + \frac{\partial f}{\partial y} dy + \frac{\partial f}{\partial z} dz = 0.$$

Now for a displacement on $f(x, y, z) = 0$, any two of dx, dy, dz may be regarded as arbitrary, and hence the condition is equivalent to

$$\frac{\partial \phi}{\partial x} \bigg/ \frac{\partial f}{\partial x} = \frac{\partial \phi}{\partial y} \bigg/ \frac{\partial f}{\partial y} = \frac{\partial \phi}{\partial z} \bigg/ \frac{\partial f}{\partial z} \tag{12}$$

when $f(x, y, z) = 0$.

General ellipsoidal forms

To demonstrate the existence of spheroidal and ellipsoidal forms it is now only necessary to show that the conditions can be satisfied for surfaces of these types. Thus, for an ellipsoidal form, by means of (10), the pressure is given by

$$\frac{p}{\rho} = \tfrac{1}{2}\omega^2(x^2 + y^2) - \pi G \rho(\alpha x^2 + \beta y^2 + \gamma z^2 - \delta) + C$$

where C is a constant. The surfaces of equal pressure may accordingly be written

$$\left(\alpha - \frac{\omega^2}{2\pi G \rho}\right) x^2 + \left(\beta - \frac{\omega^2}{2\pi G \rho}\right) y^2 + \gamma z^2 = \text{constant},$$

and for the surface $\dfrac{x^2}{a^2} + \dfrac{y^2}{b^2} + \dfrac{z^2}{c^2} = 1$ to coincide with one of these, the conditions

(12) become

$$a^2\left(\alpha - \frac{\omega^2}{2\pi G\rho}\right) = b^2\left(\beta - \frac{\omega^2}{2\pi G\rho}\right) = c^2\gamma, \tag{13}$$

and these are equivalent to

$$\left.\begin{aligned} a^2b^2(\alpha - \beta) + (a^2 - b^2)c^2\gamma &= 0 \\ \frac{\omega^2}{2\pi G\rho} = \frac{a^2\alpha - b^2\beta}{a^2 - b^2} &= \frac{b^2\beta - c^2\gamma}{b^2}. \end{aligned}\right\} \tag{14}$$

and

The first of these is a relation between the axes. The second determines (for given ρ) the appropriate value of ω in terms of the axes, or for given ω is another relation between the axes.

Substituting for β and γ from (4) in the last term gives, after a little reduction,

$$\frac{\omega^2}{2\pi G\rho} = \frac{ac}{b}(b^2 - c^2)\int_0^\infty \frac{\lambda\, d\lambda}{(b^2 + \lambda)(c^2 + \lambda)\Delta}, \tag{15}$$

which form shows that we must always have $b > c$. On the other hand it is evident from the first form for ω involved in (14), that we may have either $a \geqslant b$ or $b \geqslant a$, but it will be convenient throughout to assume always $a \geqslant b$, there being no physical difference between the two series. Thus in any case the rotation must always be about the least axis, and hence there are no equilibrium configurations having the form of prolate spheroids.

From the definitions (4) of α, β, γ we have

$$\alpha - \beta = -(a^2 - b^2)\int_0^\infty \frac{abc}{(a^2 + \lambda)(b^2 + \lambda)} \cdot \frac{d\lambda}{\Delta},$$

and hence the first of conditions (14) may be written

$$(a^2 - b^2)\int_0^\infty \left\{\frac{a^2b^2}{(a^2 + \lambda)(b^2 + \lambda)} - \frac{c^2}{c^2 + \lambda}\right\}\frac{d\lambda}{\Delta} = 0, \tag{16}$$

and this is a general relation between the axes of all ellipsoidal forms. It is clear that there may be two possible ways of satisfying it. Either, we can take $a = b$, so that the first factor vanishes; or, supposing it possible, as in fact it is, with $a \neq b$, we may make the integral factor vanish. The former solution corresponds to the Maclaurin spheroids, and the latter to the Jacobi ellipsoids. We proceed to consider in turn these two modes of satisfying the relation.

The Maclaurin spheroids

If $a = b$, the integrals for α, β, γ, and δ are expressible in terms of elementary functions, and the formula (15) for the angular velocity reduces to

$$\frac{\omega^2}{2\pi G\rho} = \frac{3 - 2e^2}{e^3}\sqrt{(1 - e^2)} \cdot \sin^{-1}e - \frac{3}{e^2}(1 - e^2), \tag{17}$$

where e is the eccentricity of the meridional sections. It will be noticed that $\omega^2/2\pi G\rho$ is independent of the size of the spheroid and is a function solely of the degree of flattening.

To calculate the angular momentum H of the system, we have

$$H = \tfrac{2}{5} M a^2 \omega,$$

and

$$M = \tfrac{4}{3} \pi \rho a^2 c.$$

If $r = (abc)^{1/3}$, then

$$\frac{H^2}{GM^3 r} = \frac{6}{25} \cdot \frac{\omega^2}{2\pi G\rho} \cdot (1 - e^2)^{-2/3}. \qquad (18)$$

The kinetic energy T at any stage is $\tfrac{1}{5} M a^2 \omega^2$, and with $GM^2 r^{-1}$ as unit of energy this may be written

$$T = \frac{3}{10} \cdot \frac{a^2}{r^2} \cdot \frac{\omega^2}{2\pi G\rho}. \qquad (19)$$

The following table gives a series of values of the various quantities for the Maclaurin spheroids.

TABLE I. MACLAURIN'S SPHEROIDS

e	a/r	c/r	$\omega^2/2\pi G\rho$	$H/G^{\frac{1}{2}}M^{\frac{3}{2}}r^{\frac{1}{2}}$	V	T
0·0	1·0	1·0	0·0	0·0	−0·6	0·0
0·1	1·0016	0·9967	0·0027	0·0255	−0·6000	0·0008
0·2	1·0068	0·9865	0·0107	0·0514	−0·6000	0·0033
0·3	1·0159	0·9691	0·0243	0·0787	−0·5999	0·0075
0·4	1·0295	0·9435	0·0436	0·1085	−0·5997	0·0139
0·5	1·0491	0·9086	0·0690	0·1417	−0·5989	0·0228
0·6	1·0772	0·8618	0·1007	0·1804	−0·5974	0·0351
0·7	1·1188	0·7990	0·1387	0·2283	−0·5941	0·0521
0·8	1·1856	0·7114	0·1816	0·2934	−0·5866	0·0766
0·8127	1·1973	0·6976	0·1868	0·3035	−0·5850	0·0805
0·9	1·3189	0·5749	0·2203	0·4000	−0·5660	0·1150
0·91	1·3411	0·5560	0·2225	0·4156	−0·5621	0·1200
0·92	1·3664	0·5355	0·2241	0·4330	−0·5575	0·1253
0·9299	1·3957	0·5134	0·2247	0·4524	−0·5520	0·1313
0·93	1·3960	0·5131	0·2247	0·4525	−0·5520	0·1314
0·94	1·4311	0·4883	0·2239	0·4748	−0·5453	0·1376
0·95	1·4740	0·4603	0·2213	0·5008	−0·5370	0·1442
0·9529	1·4884	0·4514	0·2201	0·5092	−0·5342	0·1463
0·96	1·5286	0·4280	0·2160	0·5319	−0·5262	0·1515
0·97	1·6023	0·3895	0·2063	0·5692	−0·5116	0·1588
0·98	1·7128	0·3409	0·1890	0·6249	−0·4898	0·1664
0·99	1·9210	0·2710	0·1551	0·7121	−0·4509	0·1717
0·9912	1·9622	0·2597	0·1448	0·7277	−0·4436	0·1719
1·0	∞	0·0	0·0	∞	0·0	0·0

This table is based on values given by Lamb and taken originally from Thomson and Tait, but the values of V and T have been computed by the author and also the whole of the values appertaining to the three spheroids of eccentricity 0·9299, 0·9529, and 0·9912. The first of these is the form of maximum angular velocity, the second represents the form at which ordinary stability ceases, and the last is the form possessing maximum kinetic energy.

Reference to this table makes clear the following results, which can also be rigorously established by analytical means from the foregoing theory.

(i) Every oblate spheroid is a possible equilibrium form if the angular momentum of the mass is suitably assigned.

(ii) The maximum possible value of the angular velocity is given by $\omega^2 = 2\pi G\rho \times 0·2247$ and occurs for $e = 0·9299$, $a/c = 2·7198$.

(iii) For each value of ω^2 less than this there are *two* possible spheroidal forms, but possessing different angular momentum. For values of ω^2 greater than this no spheroidal forms exist.

(iv) The angular momentum increases monotonically and indefinitely as e increases from 0 to 1, so that for given angular momentum there is only *one* possible spheroidal form.

(v) The potential energy, as it evidently must, increases monotonically with e from -0.6 for the spherical form to zero for the infinite disc, which is the limiting form of the mass for $e = 1$.

(vi) The kinetic energy begins at zero for $e = 0$ and rises to a maximum value of 0.1719 at $e = 0.9912$ and then decreases to zero for the infinite disc.

The Jacobi ellipsoids

To demonstrate the existence of ellipsoidal equilibrium forms, with three unequal axes, we consider the second way of satisfying condition (16), namely

$$\int_0^\infty \left\{ \frac{a^2 b^2}{(a^2+\lambda)(b^2+\lambda)} - \frac{c^2}{c^2+\lambda} \right\} \frac{d\lambda}{\Delta} = 0,$$

which, since $\Delta^2 = (a^2+\lambda)(b^2+\lambda)(c^2+\lambda)$, may be written

$$\int_0^\infty \{a^2 b^2 - (a^2+b^2+\lambda)c^2\} \frac{\lambda d\lambda}{\Delta^3} = 0. \tag{20}$$

If in this equation we write θa for a, θb for b, θc for c and $\theta^2 \lambda$ for λ, where θ is any constant, the equation is unaltered. Accordingly it must represent a relation between the ratios $a:b:c$ only. That is, given $a:b$ say, it determines $a:c$; or given a and b it determines c, and, as can be shown, a unique c, and then any multiples $\theta a, \theta b, \theta c$ of these values also satisfy it. It is therefore always permissible to choose the unit of length so that the product of the axes of the ellipsoid $abc = 1$.

That a solution of (20) always exists for any ratio of $a:b$ can readily be shown. For if on the left-hand side we put $c^2 = 0$, the integral remaining is essentially positive, whereas if we put $c^2 = a^2 b^2/(a^2+b^2)$ the integral is essentially negative. Since this second value of c^2 is necessarily less than both a^2 and b^2, it follows by continuity that a root must exist less than the smaller of these, and hence that there is always some value of c, less than both a and b, for which the equation is satisfied. This means that for given $a:b$, there is not only always a corresponding c, but its value leads to a real value of ω, by (15).

The angular velocity, from (14), may be put in the form

$$\frac{\omega^2}{2\pi G\rho} = abc \int_0^\infty \frac{\lambda}{(a^2+\lambda)(b^2+\lambda)} \cdot \frac{d\lambda}{\Delta}, \tag{21}$$

so that, as for the spheroids, the non-dimensional quantity $\omega^2/2\pi G\rho$ depends only on the ratios of the axes and not on their absolute lengths.

The angular momentum H of the system can be expressed in dimensionless form as follows:

$$\frac{H^2}{GM^3 r} = \frac{3}{50} \cdot \frac{\omega^2}{2\pi G\rho} \cdot \frac{(a^2+b^2)^2}{abcr}. \tag{22}$$

The potential energy V at any stage is given by (5), and the kinetic energy is now $\frac{1}{10}M(a^2+b^2)\omega^2$, which in the same units as for the spheroids and with $abc = 1$ becomes

$$\frac{3}{20}(a^2+b^2)\frac{\omega^2}{2\pi G\rho}. \tag{23}$$

For the ellipsoidal forms, the integrals involved can no longer be evaluated in terms of elementary functions. They are, however, expressible in terms of Legendre's elliptic integrals of the various kinds, as Darwin has shown, but numerical values are much more awkward to obtain than for spheroids. Their computation appears to have been tackled only by Darwin, if we except some much earlier values given by Plana, several of which Darwin considered to be wholly erroneous. Table II, calculated originally by Darwin, gives a set of values for possible ellipsoidal configurations beginning with the spheroidal member of the Jacobi series.

TABLE II. JACOBI ELLIPSOIDS

a	b	c	$\omega^2/2\pi G\rho$	H	V	T
1·197	1·197	0·698	0·187	0·304	−0·585	0·0805
1·216	1·179	0·698	0·187	0·304	(−0·585)	(0·0806)
1·279	1·123	0·696	0·186	0·306	(−0·584)	(0·0809)
1·383	1·045	0·692	0·181	0·313	−0·581	0·0817
1·601	0·924	0·677	0·166	0·341	−0·561	0·0850
1·8858	0·8150	0·6507	0·1420	0·3896	(−0·552)	(0·0894)
1·899	0·811	0·649	0·141	0·392	−0·551	0·0901
2·346	0·702	0·607	0·107	0·481	−0·519	0·0964
3·129	0·588	0·543	0·066	0·639	−0·467	0·1006
5·041	0·452	0·439	0·026	1·009	−0·355	0·0993
∞	0·0	0·0	0·0	∞	0·00	0·000

In this table the values enclosed in brackets have been interpolated from Darwin's values and not directly computed. The values in the sixth row given to an extra place of decimals correspond to the ellipsoid of bifurcation at which the pear-shaped series crosses the Jacobi series.

The table shows that for given a and b, there is only one value of c satisfying (20), a result first obtained by Meyer. It is also seen that the angular velocity is greatest for the spheroidal member and thereafter diminishes steadily as the series is described in the direction of increasing angular momentum. The figure of bifurcation common to the Maclaurin and Jacobi series can readily be found by putting $a = b$ in (20) and writing $c^2 = a^2(1-e^2)$. The integral can be evaluated in simple terms and the condition reduces to

$$\sin^{-1}e = (1-e^2)^{1/2}\frac{10e^3+3e}{3+8e^2-8e^4}, \tag{24}$$

which has for solution $e = 0·8127$.

With increasing angular momentum, b and c tend to equality, but with b always greater than c until the limiting form with infinite angular momentum which consists of an infinite cylinder with $a = \infty$, $b = c = 0$, and $abc = 1$.

As for the spheroids, the potential energy increases continually as the ellipsoid elongates with increasing angular momentum. The kinetic energy, however, rises to a maximum of about 0·1010 for a certain ellipsoid slightly more elongated than the member for which $a : b : c = 3{\cdot}129 : 0{\cdot}588 : 0{\cdot}543$.

Analytically there are, of course, two Jacobi series branching off the Maclaurin series, but they are geometrically and physically identical, and involve only an interchange of a and b. It is therefore only necessary to consider the branch for which $a \geqslant b$.

STABILITY OF THE SPHEROIDAL AND ELLIPSOIDAL FORMS FOR RESTRICTED TYPES OF DEFORMATION

Having established the existence of the Maclaurin and Jacobi series, the next question that arises is that of their stability. But we are not in a position to consider general displacements of such systems for as yet we have no means of evaluating the changes in gravitational potential energy and moment of inertia for a general deformation. If, however, we restrict the displacements simply to those for which the free surface remains a spheroid or an ellipsoid, with the least axis always coinciding with the axis of rotation, the potential energy and moment of inertia are given by expressions already available. If under such restricted conditions a system is shown to be stable, nothing more than this will have been achieved, but if on the other hand a system is shown to be unstable, then in spite of the restricted freedoms involved, the system can be regarded as unstable from a physical standpoint, and there is no strict necessity to consider any other kinds of displacement since a general displacement may be assumed to contain contributions to all possible types of deformation.

The discussion of stability in these restricted conditions that we are about to consider also provides interesting illustrations of the application of the criterions of stability developed earlier and will help to prepare for the more general discussions given in the subsequent chapters.

Stability of spheroidal forms for increasing angular velocity

Suppose we consider the Maclaurin series and take the angular velocity of the coordinate axes as the gradually increasing parameter. Also let us suppose that the deformations contemplated are such that the form always remains spheroidal with the axis of symmetry coincident with the axis of rotation. The system then has only one degree of freedom and the single coordinate required for its specification may be taken as e the eccentricity of the sections through the axis of rotation. This example is, of course, artificial from a physical standpoint, since angular velocity is not the physically significant factor in a free system.

The stability or otherwise of the system then depends on the total mechanical potential

$$U = V - \tfrac{1}{2}\omega^2 I,$$

where V is given by (8), and the moment of inertia by

$$I = \tfrac{2}{5}Ma^2 = \tfrac{2}{5}Mr^2(1-e^2)^{-1/3}.$$

We thus have

$$U = -\tfrac{16}{15}\pi^2 G\rho^2 r^5 \left\{ (1-e^2)^{1/6} \frac{\sin^{-1}e}{e} + 2k(1-e^2)^{-1/3} \right\}$$

where $k = \omega^2/2\pi G\rho$ and remains constant during the deformations. In the present example ω will not be the angular velocity of the spheroid at all times during the deformations, but only of the equilibrium form in the neighbourhood of which the motion is taking place. It is of course the angular momentum in a fixed frame that remains constant. It will be convenient to use e to represent the eccentricity during the deformations and also the eccentricity of the particular member of the Maclaurin series about which the motion is supposed to occur.

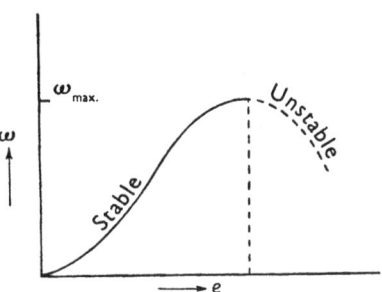

Fig. 10. Graph of ω plotted against e for Maclaurin spheroids.

Omitting the constant factor $\tfrac{16}{15}\pi^2 G\rho^2 r^5$, equilibrium configurations will be given by the vanishing of the derivative, thus

$$-\frac{dU}{de} = (1-e^2)^{-5/6}\left(\frac{2}{3}-\frac{1}{e^2}\right)\sin^{-1}e + \frac{1}{e}(1-e^2)^{-1/3} + \tfrac{1}{3}ke(1-e^2)^{-4/3} = 0,$$

and this equation is satisfied, as it should be, when k has the value given by (17), so that the Maclaurin spheroids are the equilibrium forms. To examine their stability, we have, omitting positive constant factors,

$$-\frac{d^2U}{de^2} = (1-e^2)^{-11/6}\sin^{-1}e\left(-\frac{8}{3e}+\frac{3}{e^3}\right) - (1-e^2)^{-4/3}\left(\frac{3}{e^2}-\frac{2}{3}\right).$$

It may readily be shown from this expression that $\dfrac{d^2U}{de^2}$ vanishes for $e=0$, but immediately thereafter is positive and remains so until that value of e given by

$$\sin^{-1}e = (1-e^2)^{1/2}(9e-2e^3)/(9-8e^3)$$

at which it changes sign and then remains negative. It can easily be verified from (17) that this value of e corresponds precisely to that value at which $\dfrac{d\omega}{de} = 0$. Accordingly stability, under the present conditions, disappears, as would be expected, at the maximum value of the angular velocity. It will be shown later, however, that this is not a true instability. In fact, the result means only that relative to axes rotating at a rate greater than $\omega_{\text{max.}}$, no spheroidal form of equilibrium is possible, and the coordinates in any steady state if measured in such a frame would appear to depart more and more from any given initial values.

Stability of the spheroids for increasing angular momentum

If the angular momentum H is adopted as parameter and the deformations again restricted to those maintaining the spheroidal form, the Maclaurin series can readily be shown to be thoroughly stable, and the preceding result accordingly shown to be a spurious instability arising purely from the choice of axes.

In the present case the appropriate function for discussing secular stability is

$$U = V + H^2/2I,$$

where $\qquad H = \tfrac{2}{5}Ma^2\omega \quad \text{and} \quad I = \tfrac{2}{5}Ma^2 = \tfrac{2}{5}Mr^2(1-e^2)^{-1/3}.$

The displacements at any stage are represented by varying e but with H remaining constant. We thus have

$$U = -\frac{16}{15}\pi^2 G\rho^2 r^5 (1-e^2)^{1/6}\frac{\sin^{-1}e}{e} + \frac{5H^2}{4Mr^2}(1-e^2)^{1/3}$$

$$= -\frac{16}{15}\pi^2 G\rho^2 r^5 \left\{ (1-e^2)^{1/6}\frac{\sin^{-1}e}{e} - \kappa(1-e^2)^{1/3} \right\},$$

where κ is a factor that remains constant during the deformations and is given by

$$\kappa = \frac{5H^2}{4Mr^2}\cdot\frac{15}{16\pi^2 G\rho^2 r^5}$$

$$= \frac{\omega^2}{2\pi G\rho}\cdot\frac{1}{2(1-e^2)^{2/3}}.$$

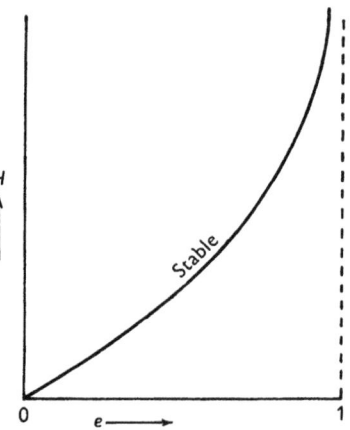

H

Fig. 11. Graph of H plotted against e for Maclaurin spheroids.

Omitting the constant factor $\frac{16}{15}\pi^2 G\rho^2 r^5$, the equilibrium forms are, as usual, given by

$$-\frac{dU}{de} = (1-e^2)^{-5/6}\sin^{-1}e\left(\frac{2}{3}-\frac{1}{e^2}\right) + e^{-1}(1-e^2)^{-1/3} + \tfrac{2}{3}\kappa e(1-e^2)^{-2/3} = 0$$

and, as before, this equation is satisfied for the value of κ given above. To examine the question of stability, we have

$$-\frac{d^2U}{de^2} = (1-e^2)^{-11/6}\sin^{-1}e\left(\frac{8}{9}e - \frac{4}{e} + \frac{3}{e^3}\right) - (1-e^2)^{-4/3}\left(\frac{3}{e^2}-2\right).$$

From this it may easily be shown that $\dfrac{d^2U}{de^2}$ vanishes only for $e = 0$ and that thereafter it is always positive. It follows that every member of the Maclaurin series is secularly stable for displacements that preserve the spheroidal form with axis of symmetry remaining the axis of angular momentum.

The foregoing results concerning the stability of the spheroids could have been deduced at once from the considerations of Chapter II. Thus, when ω is taken as parameter, we see from Fig. 10 that at a certain value of e the derivative $\dfrac{d\omega}{de} = 0$, so that stability, in the coordinate system implied, is lost. On the other hand, if H is taken as parameter, it increases monotonically with e, and there is no configuration at which $\dfrac{dH}{de} = 0$. Since the spherical form is stable, the series must therefore be stable everywhere.

Stability of spheroids and ellipsoids for certain ellipsoidal deformations

If the displacements are assumed to be such as to preserve always an ellipsoidal form with the axis of rotation a principal axis, the question of secular stability can again be dealt with quite simply.

If we plot the values of a/c and b/c for the Maclaurin and Jacobi series, determined by Tables I and II, as rectangular coordinates, we get the diagram of Fig. 12. In this diagram the point S corresponds to the spherical form ($a = b = c$),

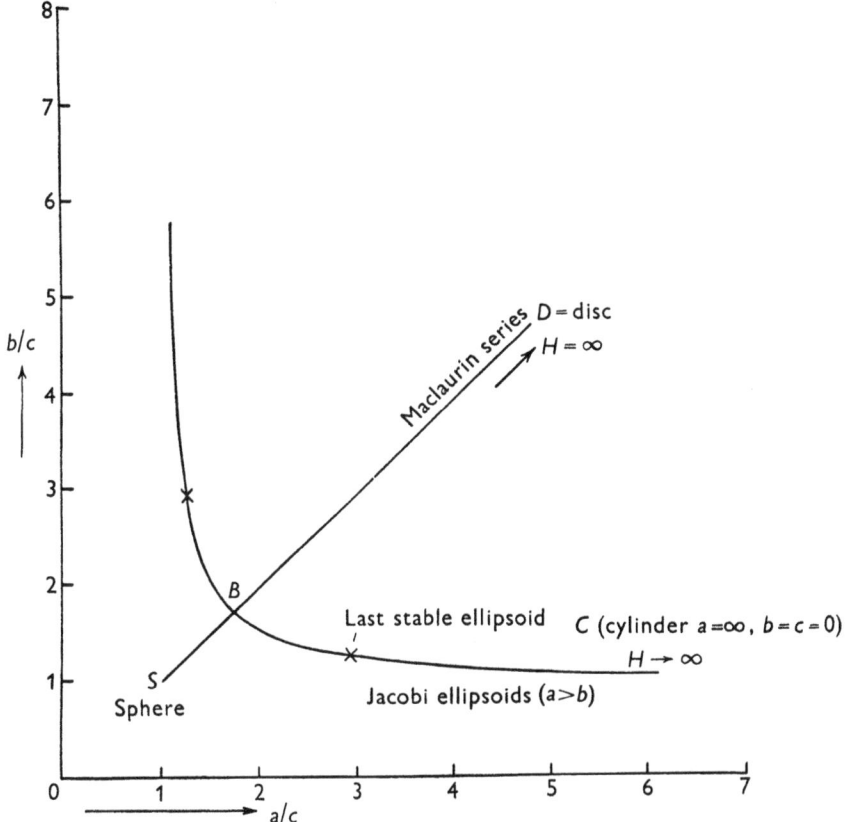

Fig. 12. Diagram showing representative points of Jacobi and Maclaurin figures.

D corresponds to the infinite thin disc-spheroid ($a=b=\infty$, $c=0$), and C corresponds to the infinitely elongated cylindrical ellipsoid ($a=\infty$, $b=c=0$), which is the limiting form of the Jacobi figure. The point B is the point of bifurcation of the two linear series.

It is important to notice that we are considering a system defined by three unequal axes and so requiring two coordinates for its specification. If we start with the Maclaurin spheroids we are limited to the case $a = b$, which requires only a single coordinate, and the existence of the Jacobi ellipsoids does not manifest itself. This is why no maximum or critical value of the angular momentum, corresponding to the point B, occurs in Table I for the Maclaurin forms. On the other hand, the value of H at the point B for the Jacobi series has a critical value. Thus the spheroids can be regarded as a special case of ellipsoidal forms, so that there are two series of ellipsoidal forms intersecting at B, namely, the Maclaurin and Jacobi series. But there is only one series of spheroidal forms.

To discuss their stability by diagrammatic means we can imagine the linear series of Jacobi forms described by beginning with $H = \infty$ and the angular momentum gradually diminished, that is, we can use H as a decreasing parameter. Then, at first, for sufficiently large H, there are three possible equilibrium forms (two equivalent ellipsoids and one spheroid), but the parameter eventually reaches a value where $dH = 0$, that is, at B. At this stage the forms with three unequal axes disappear, and for smaller H only the spheroidal form is possible. The system thus provides an instance of case (i) of Chapter II (p. 10). Since the spherical form is stable it follows that the Maclaurin series is stable for the portion SB of the series and that it loses its stability and is always unstable thereafter. Moreover, since the Jacobi series then comes into existence for greater values of H, stability will be transferred to it. Plotted in terms of angular momentum as vertical coordinate the Jacobi series turns upwards.

On the other hand, if the angular velocity were used as parameter and plotted as vertical coordinate, it is clear from Tables I and II that the ellipsoidal series would turn downwards after B. If the system were regarded as evolving along the Maclaurin series SB, with increasing angular velocity, the ellipsoidal forms could not come into existence, and in this sense the Jacobi series is unstable.

Analytical proof of stability for ellipsoidal displacements

The foregoing discussion of secular stability for restricted deformations is based on arguments given by Jeans. A simple analytical proof, under the same restrictions as to the permitted deformations, has been given by Lamb. The general lines of his discussion appear to be sound, but it is incomplete in that certain points necessary to the argument are left unestablished. These points are covered in the proof now to be given.

For secular stability, as the angular momentum is increased, the function $V + H^2/2I = f(a, b)$, say, must be an absolute minimum for ellipsoidal deformations in which the angular momentum remains constant. We have from (5)

$$f(a, b) = -\tfrac{8}{15}\pi^2 G\rho^2 r^6 \int_0^\infty \frac{d\lambda}{\sqrt{[(a^2 + \lambda)(b^2 + \lambda)(c^2 + \lambda)]}} + \frac{5H^2}{2M(a^2 + b^2)},$$

and since $abc = r^3$, this expression, as indicated, may be regarded as a function of a and b only.

Whatever value H may have, this function has the following values at the points indicated:

$$\text{for} \quad a = \infty, \quad f(\infty, b) = 0,$$
$$\text{for} \quad b = \infty, \quad f(a, \infty) = 0,$$
$$\text{for} \quad a = b = 0, \quad f(0, 0) = \infty.$$

Moreover, $f(a, b)$ must always exceed the finite negative value of the first term corresponding to the spherical form.

Also, if $a = 0$, so that $bc = \infty$, $\qquad \int_0^\infty \dfrac{d\lambda}{\Delta} \to 0,$

and hence $\qquad\qquad\qquad\qquad f(0, b) = \dfrac{5H^2}{2M} . b^{-2},$

and similarly, if $b = 0$, $\qquad\qquad f(a, 0) = \dfrac{5H^2}{2M} . a^{-2}.$

It is also necessary to the argument to show that as $a^2 + b^2 \to \infty$, $f(a, b)$ tends to zero through negative values provided neither a nor b vanishes. To prove this, let $a = r\epsilon^{-1}$, $b = r\epsilon^{-1+s}$, so that $c = r\epsilon^{2-s}$, where $0 < s < 1$, so that $c \to 0$ and $a, b \to \infty$, as $\epsilon \to 0$. Then the gravitational term, omitting the constant factor, becomes

$$-\int_0^\infty \frac{d\lambda}{\{(r^2 \epsilon^{-2} + \lambda)(r^2 \epsilon^{-2+2s} + \lambda)(r^2 \epsilon^{4-2s} + \lambda)\}^{1/2}}$$

$$= - \underset{K \to \infty}{\mathrm{Lt}} \int_0^K \frac{\epsilon^{2-s} d\lambda}{\{(r^2 + \epsilon^2 \lambda)(r^2 + \epsilon^{2-2s} \lambda)(r^2 \epsilon^{4-2s} + \lambda)\}^{1/2}}.$$

Now suppose that $K = \epsilon^{-k}$, where $k > 0$, so that $K \to \infty$ as $\epsilon \to 0$, and that $K\epsilon^2 \to 0$ and $K\epsilon^{2-2s} \to 0$. These require $k < 2$ and $0 < 2 - 2s - k$. The terms in the denominator involving ϵ can then be ignored, and the integral becomes

$$- \mathrm{Lt} \int_0^K \frac{\epsilon^{2-s} d\lambda}{r^2 \lambda^{\frac{1}{2}}} = - \mathrm{Lt}\, 2\epsilon^{2-s} K^{\frac{1}{2}} r^{-2} = - O(\epsilon^{2-s-\frac{1}{2}k}).$$

The second term in $f(a, b)$ is clearly $O(\epsilon^2)$. Hence the negative gravitational term will predominate at infinity provided

$$2 - s - \tfrac{1}{2}k < 2.$$

Since this is certainly true for $s \geqslant 0$, it follows that the function approaches the infinite circle from below except at the points on the axes $a = 0$, $b = 0$. The secular stability can now be readily established as follows.

If (i) $H < 0\cdot304\, G^{\frac{1}{2}} M^{\frac{5}{6}} r^{\frac{1}{6}} = H_c$, say, then Tables I and II show that there is only one possible equilibrium form and that this is a Maclaurin spheroid. Hence a critical value of $f(a, b)$ must occur for a certain point such that $a = b$. If therefore a, b, and f are plotted as rectangular coordinates with the a, b axes horizontal and f vertical, the resulting surface (for $H < H_c$) must be of the form shown in Fig. 13. Since $f(0, 0) = +\infty$ and $f(\infty, \infty) = -0$, the critical point M in the plane $a = b$ corresponds to a negative value of f and must be an absolute minimum with respect to all variations on the surface, as otherwise there would have to be further critical points on the surface. The Maclaurin spheroids for $H < H_c$ are accordingly secularly stable for displacements of the kind permitted.

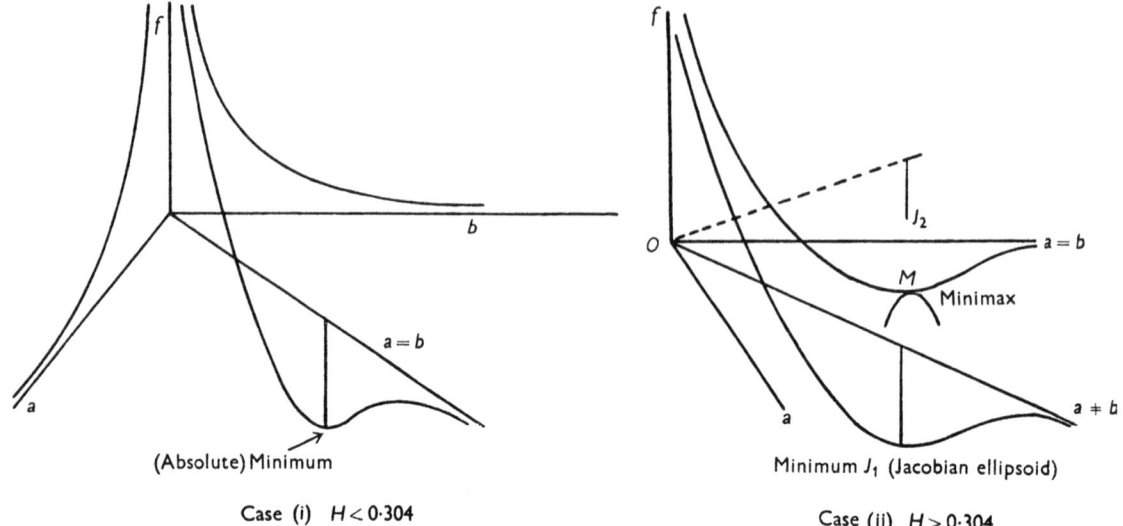

(Absolute) Minimum

Case (i) $H < 0.304$

Minimum J_1 (Jacobian ellipsoid)

Case (ii) $H > 0.304$

Fig. 13. Case (i) $H < 0.304$. Case (ii) $H > 0.304$. The point M in the plane $a = b$, representing a Maclaurin form is a minimum for displacements in this plane but a maximum for displacements perpendicular to it.

$J_1(a \neq b)$, an absolute minimum, and a second symmetrical point J_2, give together the two *stable* Jacobi forms.

If (ii) $H > 0.304 \, G^{\frac{1}{4}} M^{\frac{5}{6}} r^{\frac{1}{4}}$, Tables I and II show that there are three critical values of f, one of which always corresponds to a spheroid and is therefore in the plane $a = b$, and the remaining two correspond to equivalent Jacobi ellipsoids, with a and b interchanged, and therefore symmetrically placed with regard to the plane $a = b$. Let us denote these critical points by J_1 and J_2. The plane through J_1, say, and the vertical axis meets the surface in a certain curve, the ordinate of which is $+\infty$ for $a = b = 0$ and -0 at the infinite circle $a^2 + b^2 = \infty$, $f = 0$. The critical value at J_1 is therefore negative and finite, and it follows that J_1 and J_2 must be points where f is an absolute minimum. Also the point M must now be a so-called minimax, since it is a minimum for displacements in the plane $a = b$ but a maximum for displacements transverse to this. This arrangement of critical points is the only possibility consistent with there being three, and only three, stationary points on the surface.

It can therefore be concluded that the Maclaurin spheroids are secularly stable for *ellipsoidal* displacements of the present kind for $e < 0.8127$, though this by no means enables us to assert their stability for general deformations. On the other hand, their secular instability for $e > 0.8127$, which the foregoing argument establishes, means that they may be considered physically unstable beyond this limit, and that they would not therefore come into existence as a result of an evolutionary process that could be represented by increasing angular momentum. It will be shown later that the spheroids are secularly stable for all displacements if $e < 0.8127$.

Where ordinary stability is concerned, it follows that if the oscillations are restricted to forms of the present ellipsoidal type, the spheroids are ordinarily stable for $e < 0.8127$, but for values greater than this nothing can be inferred from the present work as to the ordinary stability. It has, however, been shown by Cartan that they remain ordinarily stable for all displacements provided $e < 0.9529$. But physically the spheroids are of little interest beyond the point where their secular stability ceases.

For the Jacobi forms the argument has shown that they are always stable for these restricted ellipsoidal deformations, but nothing can be inferred from this result concerning their stability, or otherwise, for general displacements. It will prove in the sequel that the Jacobi forms remain stable for all displacements up to the configuration for which

$$a : b : c : (abc)^{1/3} = 1.8858 : 0.8150 : 0.6507 : 1$$

and
$$H = 0.3896\, G^{\frac{1}{2}}\, M^{\frac{2}{3}}\, r^{\frac{1}{2}}.$$

But the deformation through which instability first enters is only disclosed when the displacement is analysed in general form by means of ellipsoidal harmonics.

Evolution along a linear series

In the foregoing discussions the angular momentum has usually been adopted as the gradually changing parameter, while the density of the mass is assumed to remain constant. It can be shown, however, that the series of configurations obtained is exactly the same, as to its proportions, if the density gradually increases while the angular momentum remains fixed. To see this, since the total mass and angular momentum are given by

$$M = \tfrac{4}{3}\pi \rho abc, \quad H = \tfrac{1}{5}M(a^2 + b^2)\omega,$$

there exists the following relation involving only ratios of lengths

$$H\rho^{2/3} = 10\left(\frac{2\pi^2}{9}\right)^{1/3} M^{5/3} \frac{a^2 + b^2}{(abc)^{2/3}}\, \omega.$$

Suppose now that the total mass M, the angular velocity ω, and the ratios of the axes are fixed, then the value of the right-hand side is fixed, and the relation can therefore be satisfied for any assigned ρ by appropriately choosing H, or for any assigned H by appropriately choosing ρ. It follows that the linear series given in Tables I and II obtained by increasing H could equally well be obtained by increasing ρ with H remaining constant. Since, when ρ is maintained constant, we can assume $abc = 1$ always, this usually makes it more convenient to regard the series as described by increasing H with the density remaining constant.

When the angular momentum is kept constant, the initial spherical member of the Maclaurin series will correspond to zero angular velocity, zero density, and infinite radius, while the final members of both the Maclaurin and Jacobi series will have infinite density. It is easily shown from Tables I and II that when H is kept constant the angular velocity increases always with the density. To determine the values of ω and of ρ for any given form, that is, for given ratios of the axes as already tabulated, it is necessary, since ρ is no longer constant along the series, to choose some unit of density. For this it is convenient to select that value corresponding to the figure common to the Maclaurin and Jacobi series, and to choose the unit of angular momentum so that the constant value of H is 0·3035. It may further be supposed that the radius which the bifurcation figure would possess if changed to spherical form without change of density is the unit of length.

Since the dimensions of H are the same as of $G^{\frac{1}{2}} M^{\frac{2}{3}} r^{\frac{1}{2}}$, it follows that for any other configuration, of given ratios of axes, the value of its spherical radius r must be inversely proportional to the square of the tabulated angular momentum, and hence that the density must be proportional to the sixth power of this quantity. Thus, to take particular examples, for the figure of bifurcation we have

$$a : b : c = 1{\cdot}197 : 1{\cdot}197 : 0{\cdot}698,$$

$$\omega^2/2\pi G = 0{\cdot}1871, \quad H = 0{\cdot}3035,$$

and hence for the critical Jacobi figure for which

$$a : b : c = 1{\cdot}886 : 0{\cdot}815 : 0{\cdot}651 \quad \text{and} \quad H \text{ (tabulated)} = 0{\cdot}3896$$

we have $\quad \omega^2/2\pi G = 0{\cdot}142 \times (0{\cdot}3896/0{\cdot}3035)^6 = 0{\cdot}636, \quad (H = 0{\cdot}3035).$

Again, for the last but one tabulated member of the Jacobi series for which

$$a : b : c = 3{\cdot}129 : 0{\cdot}588 : 0{\cdot}543 \quad \text{and} \quad H \text{ (tabulated)} = 0{\cdot}639$$

we have $\quad \omega^2/2\pi G = 0{\cdot}066 \times (0{\cdot}639/0{\cdot}3035)^6 = 5{\cdot}75, \quad (H = 0{\cdot}3035).$

In the case of uniform density, ω decreases to zero as the angular momentum increases to infinity and the kinetic energy of the motion has a maximum value after which it decreases to zero for the infinitely elongated form. Since the kinetic energy must also be zero in the infinitely dense final form, when the density is used as parameter, it must again have a maximum value for some density intermediate between 0 and ∞. These results appear to have been first established by G. H. Darwin.

Chapter V

ELLIPSOIDAL HARMONIC ANALYSIS

In the present chapter we develop the elements of ellipsoidal harmonic analysis as far as is necessary in establishing those properties that are required for the subsequent chapters. Standard results in the theory of spherical harmonics will be assumed.

Confocal coordinates

Consider the system of quadrics represented in rectangular cartesian coordinates x, y, z, by

$$\frac{x^2}{a^2+\theta}+\frac{y^2}{b^2+\theta}+\frac{z^2}{c^2+\theta} = 1, \tag{1}$$

where θ is a variable parameter, and for which it will be supposed that $a > b > c$. They have the simple property that through any point pass three different members of the system. If x, y, z are fixed, the equation can be interpreted as a cubic in θ whose roots are the parameters of these three surfaces. This equation may be written

$$f(\theta) \equiv (a^2+\theta)(b^2+\theta)(c^2+\theta) - x^2(b^2+\theta)(c^2+\theta)$$
$$- y^2(c^2+\theta)(a^2+\theta) - z^2(a^2+\theta)(b^2+\theta) = 0. \tag{2}$$

To locate the roots, we have that $f(-a^2)$ is negative, $f(-b^2)$ is positive, $f(-c^2)$ is negative, and $f(+\infty)$ is positive. Hence the roots of the cubic are all real and different. If we denote them by λ, μ, ν, taken always so that $\lambda > \mu > \nu$, then

$$\lambda > -c^2 > \mu > -b^2 > \nu > -a^2. \tag{3}$$

It follows that the λ-surfaces are ellipsoids, the μ-surfaces are hyperboloids of one sheet, and the ν-surfaces hyperboloids of two sheets. It is also easily shown that these three families of surfaces are everywhere mutually orthogonal.

It is evident from symmetry that three given surfaces

$$\lambda = \text{constant}, \quad \mu = \text{constant}, \quad \nu = \text{constant},$$

will intersect in eight points, one in each octant defined by the coordinate planes, whose coordinates can be written $(\pm x, \pm y, \pm z)$. In discussing the properties of the surfaces it is therefore sufficient to consider simply the positive octant.

Returning to the original equation (1), if λ, μ, ν are the parameters of the three surfaces through the point (x, y, z), then we must have as an identity

$$\frac{x^2}{a^2+\theta}+\frac{y^2}{b^2+\theta}+\frac{z^2}{c^2+\theta} - 1 \equiv \frac{(\lambda-\theta)(\mu-\theta)(\nu-\theta)}{(a^2+\theta)(b^2+\theta)(c^2+\theta)}. \tag{4}$$

Multiplying both sides by $a^2+\theta$ and then putting $\theta = -a^2$ (and so on in turn) we get the following expressions for the rectangular coordinates (x, y, z) in terms of the orthogonal curvilinear coordinates (λ, μ, ν):

$$\left.\begin{aligned}
x^2 &= \frac{(\lambda+a^2)\,(\mu+a^2)\,(\nu+a^2)}{(a^2-b^2)\,(a^2-c^2)}, \\[2mm]
y^2 &= -\frac{(\lambda+b^2)\,(\mu+b^2)\,(\nu+b^2)}{(a^2-b^2)\,(b^2-c^2)}, \\[2mm]
z^2 &= \frac{(\lambda+c^2)\,(\mu+c^2)\,(\nu+c^2)}{(a^2-c^2)\,(b^2-c^2)}.
\end{aligned}\right\} \tag{5}$$

By direct addition of these, or more simply from consideration of the sum of the roots of $f(\theta) = 0$, we have

$$r^2 = x^2+y^2+z^2 = \lambda+\mu+\nu+a^2+b^2+c^2. \tag{6}$$

Since for given finite values of a, b, c the quantities μ and ν always lie within finite limits, by (3), it is evident that at great distance from the origin the parameter λ must tend to infinity. That is

$$r \to \lambda^{\frac12} \to \infty.$$

Lamé functions and Lamé polynomials

Let us suppose next that $f(t)$ denotes a rational integral polynomial of degree m in the variable t, and let us write

$$L = f(\lambda), \quad M = f(\mu), \quad N = f(\nu).$$

Then the product LMN will be a polynomial symmetrical with regard to λ, μ, and ν.

If the factors of $f(t)$ are $t-\delta_1, t-\delta_2, \ldots, t-\delta_m$ say, assuming the coefficient of t^m in $f(t)$ to be unity, we shall have

$$L = f(\lambda) = \prod_{r=1}^{m} (\lambda-\delta_r),$$

and hence that

$$LMN = \prod_{r=1}^{m} \{(\lambda-\delta_r)\,(\mu-\delta_r)\,(\nu-\delta_r)\}$$

$$= \prod_{r=1}^{m} \left\{\frac{x^2}{a^2+\delta_r}+\frac{y^2}{b^2+\delta_r}+\frac{z^2}{c^2+\delta_r}-1\right\} \tag{7}$$

by means of (4), apart from a factor independent of x, y, z. Hence LMN defined in this way will always be equal to a product of quadratic factors $Q_1, Q_2, \ldots, Q_m$ each of the form

$$Q_r = \frac{x^2}{a^2+\delta_r}+\frac{y^2}{b^2+\delta_r}+\frac{z^2}{c^2+\delta_r}-1. \tag{8}$$

If, as will be shown later is possible, the function (7) so resulting, which we may denote by $V(x, y, z)$, is such that it satisfies Laplace's equation

$$\nabla^2 V \equiv \frac{\partial^2 V}{\partial x^2}+\frac{\partial^2 V}{\partial y^2}+\frac{\partial^2 V}{\partial z^2} = 0,$$

then the original function $L(\lambda)$ is called a *Lamé function of the first kind*, or *of the first species*.

If similarly we commence by putting

(i) $L = f(\lambda)\sqrt{(\lambda+a^2)}, \quad M = f(\mu)\sqrt{(\mu+a^2)}, \quad N = f(\nu)\sqrt{(\nu+a^2)};$

or   (ii) $L = f(\lambda)\sqrt{(\lambda+b^2)}, \quad M = f(\mu)\sqrt{(\mu+b^2)}, \quad N = f(\nu)\sqrt{(\nu+b^2)};$

or   (iii) $L = f(\lambda)\sqrt{(\lambda+c^2)}, \quad M = f(\mu)\sqrt{(\mu+c^2)}, \quad N = f(\nu)\sqrt{(\nu+c^2)};$

where in each case $f(\lambda)$ is again a rational integral polynomial whose term of highest degree is λ^m, then in exactly the same way it is seen from (5) that the product LMN will be a symmetrical polynomial in λ, μ, ν multiplied in case (i) by x, in case (ii) by y, and in case (iii) by z. Moreover, the rational part of the product LMN will again be expressible as a product of quadratic factors $Q_r(x, y, z)$, so that the whole expression LMN when put in terms of x, y, z will be of one of the forms

$$V = (x, y, z) \prod_{r=1}^{m} \left\{ \frac{x^2}{a^2 + \delta_r} + \frac{y^2}{b^2 + \delta_r} + \frac{z^2}{c^2 + \delta_r} - 1 \right\}.$$

If V resulting in this way satisfies Laplace's equation, the corresponding function $L(\lambda)$, which now contains *one* of the radicals $\sqrt{(\lambda+a^2)}$, $\sqrt{(\lambda+b^2)}$, or $\sqrt{(\lambda+c^2)}$ as a factor, is called a *Lamé function of the second kind*.

Similarly, if we put

(i)

$L = f(\lambda)\sqrt{[(\lambda+b^2)(\lambda+c^2)]}, \quad M = f(\mu)\sqrt{[(\mu+b^2)(\mu+c^2)]}, \quad N = f(\nu)\sqrt{[(\nu+b^2)(\nu+c^2)]};$

or (ii)

$L = f(\lambda)\sqrt{[(\lambda+c^2)(\lambda+a^2)]}, \quad M = f(\mu)\sqrt{[(\mu+c^2)(\mu+a^2)]}, \quad N = f(\nu)\sqrt{[(\nu+c^2)(\nu+a^2)]};$

or (iii)

$L = f(\lambda)\sqrt{[(\lambda+a^2)(\lambda+b^2)]}, \quad M = f(\mu)\sqrt{[(\mu+a^2)(\mu+b^2)]}, \quad N = f(\nu)\sqrt{[(\nu+a^2)(\nu+b^2)]};$

where again $f(\lambda)$ is a rational integral polynomial, we arrive by means of the product LMN at a function of one of the forms

$$V = (yz, zx, xy) \prod_{r=1}^{m} \left\{ \frac{x^2}{a^2 + \delta_r} + \frac{y^2}{b^2 + \delta_r} + \frac{z^2}{c^2 + \delta_r} - 1 \right\},$$

and if $\nabla^2 V = 0$, then the original function $L(\lambda)$ is called a *Lamé function of the third kind*.

Finally, if

$$L = f(\lambda)\sqrt{[(\lambda+a^2)(\lambda+b^2)(\lambda+c^2)]}, \quad M = f(\mu)\sqrt{[(\mu+a^2)(\mu+b^2)(\mu+c^2)]},$$

$$N = f(\nu)\sqrt{[(\nu+a^2)(\nu+b^2)(\nu+c^2)]},$$

the resulting function V will be of the form

$$V = xyz \prod_{1}^{m} \left\{ \frac{x^2}{a^2 + \delta_r} + \frac{y^2}{b^2 + \delta_r} + \frac{z^2}{c^2 + \delta_r} - 1 \right\},$$

and if $\nabla^2 V = 0$, then the original function $L(\lambda)$, which now contains

$$\sqrt{[(\lambda + a^2)(\lambda + b^2)(\lambda + c^2)]}$$

as a factor, is termed a *Lamé function of the fourth kind.*

The several cases may be combined schematically as follows:

$$LMN = \left\{ \begin{array}{ccc} & x & yz \\ 1 & y & zx \quad xyz \\ & z & xy \end{array} \right\} \prod_1^m \left\{ \frac{x^2}{a^2 + \delta_r} + \frac{y^2}{b^2 + \delta_r} + \frac{z^2}{c^2 + \delta_r} - 1 \right\},$$

$$= V(x, y, z). \tag{9}$$

When $\nabla^2 V = 0$, the polynomial $V(x, y, z)$ is called a *Lamé polynomial* or an *ellipsoidal harmonic*, and to each species of Lamé functions there corresponds a species of ellipsoidal harmonics.

The possible forms that the Lamé functions $L(\lambda)$ can take must depend ultimately on the requirement that V must satisfy Laplace's equation, and we proceed to consider how suitable functions $L(\lambda)$ may be constructed.

Laplace's equation in confocal coordinates

Since the surfaces λ, μ, ν are mutually orthogonal, the square of the line element ds corresponding to a displacement $d\lambda, d\mu, d\nu$ will be of the form

$$ds^2 = h_1^2 d\lambda^2 + h_2^2 d\mu^2 + h_3^2 d\nu^2,$$

and it is easily found that

$$h_1^2 = \frac{(\lambda - \mu)(\lambda - \nu)}{4(a^2 + \lambda)(b^2 + \lambda)(c^2 + \lambda)},$$

$$h_2^2 = \frac{(\mu - \nu)(\mu - \lambda)}{4(a^2 + \mu)(b^2 + \mu)(c^2 + \mu)}, \tag{10}$$

$$h_3^2 = \frac{(\nu - \lambda)(\nu - \mu)}{4(a^2 + \nu)(b^2 + \nu)(c^2 + \nu)}.$$

If we define positive quantities R, A, B, and C as follows:

$$R = \sqrt{[(\lambda - \mu)(\lambda - \nu)(\mu - \nu)]}, \tag{11}$$

$$A = \sqrt{[(a^2 + \lambda)(b^2 + \lambda)(c^2 + \lambda)]},$$
$$B = \sqrt{[-(a^2 + \mu)(b^2 + \mu)(c^2 + \mu)]}, \tag{12}$$
$$C = \sqrt{[(a^2 + \nu)(b^2 + \nu)(c^2 + \nu)]},$$

then $\qquad h_1 = \dfrac{R}{2A\sqrt{(\mu - \nu)}}, \qquad h_2 = \dfrac{R}{2B\sqrt{(\lambda - \nu)}}, \qquad h_3 = \dfrac{R}{2C\sqrt{(\lambda - \mu)}}. \tag{13}$

Expressed in general orthogonal curvilinear coordinates Laplace's equation is well known to become

$$\frac{\partial}{\partial \lambda}\left(\frac{h_2 h_3}{h_1}\frac{\partial V}{\partial \lambda}\right) + \frac{\partial}{\partial \mu}\left(\frac{h_3 h_1}{h_2}\frac{\partial V}{\partial \mu}\right) + \frac{\partial}{\partial \nu}\left(\frac{h_1 h_2}{h_3}\frac{\partial V}{\partial \nu}\right) = 0,$$

and in the present case, since A, B, C depend only on λ, μ, and ν respectively, this may be written on multiplying through by ABC

$$(\mu-\nu)A\frac{\partial}{\partial\lambda}\left(A\frac{\partial V}{\partial\lambda}\right)+(\nu-\lambda)B\frac{\partial}{\partial\mu}\left(B\frac{\partial V}{\partial\mu}\right)+(\lambda-\mu)C\frac{\partial}{\partial\nu}\left(C\frac{\partial V}{\partial\nu}\right)=0,$$

or
$$\sum_{a,b,c}\left\{A^2\frac{\partial^2 V}{\partial\lambda^2}+A\frac{dA}{d\lambda}\cdot\frac{\partial V}{\partial\lambda}\right\}(\mu-\nu)=0. \qquad (14)$$

Normal solutions of Laplace's equation

Suppose that we seek solutions of equation (14) of the form
$$V = L(\lambda)M(\mu)N(\nu),$$

wherein, as indicated by the notation, L is a function of λ only, M a function of μ only, and N a function of ν only. Solutions so expressible are sometimes termed *normal solutions* of the equation. For such solutions we have

$$A\frac{\partial}{\partial\lambda}\left(A\frac{\partial V}{\partial\lambda}\right)=LMN\cdot\frac{A}{L}\frac{d}{d\lambda}\left(A\frac{dL}{d\lambda}\right),$$

and accordingly Laplace's equation becomes

$$(\mu-\nu)\frac{A}{L}\frac{d}{d\lambda}\left(A\frac{dL}{d\lambda}\right)+(\nu-\lambda)\frac{B}{M}\frac{d}{d\mu}\left(B\frac{dM}{d\mu}\right)+(\lambda-\mu)\frac{C}{N}\frac{d}{d\nu}\left(C\frac{dN}{d\nu}\right)=0,$$

which is of the form
$$(\mu-\nu)\phi(\lambda)+(\nu-\lambda)\phi_2(\mu)+(\lambda-\mu)\phi_3(\nu)=0.$$

This equation, if V is a solution of $\nabla^2 V=0$, must be true for all λ,μ,ν. Putting $\mu=\nu$ gives at once $\phi_2(\mu)=\phi_3(\mu)$, and thus all three functions ϕ, ϕ_2, and ϕ_3 must be identical, and such that

$$(\mu-\nu)\phi(\lambda)+(\nu-\lambda)\phi(\mu)+(\lambda-\mu)\phi(\nu)=0.$$

To find ϕ, if λ is put zero, this equation can be written

$$\frac{\phi(\mu)-\phi(0)}{\mu}=\frac{\phi(\nu)-\phi(0)}{\nu},$$

and since this must hold for all μ and ν, each side must in fact be constant. The form of the function ϕ is thus

$$\phi(\lambda)=H\lambda+K,$$

where H and K are constants, and the differential equation for L is accordingly

$$A\frac{d}{d\lambda}\left(A\frac{dL}{d\lambda}\right)=(H\lambda+K)L. \qquad (15)$$

For reasons that will be apparent in due course it is convenient to replace H by a different constant, n, related to it as follows

$$\tfrac{1}{4}n(n+1)=H,$$

and it will be seen later that the important cases are when n is an integer. The equation for L then becomes

$$A\frac{d}{d\lambda}\left(A\frac{dL}{d\lambda}\right)-\{\tfrac{1}{4}n(n+1)\lambda+K\}L=0. \qquad (16)$$

This linear differential equation for L is known as *Lamé's equation*.

The functions M and N satisfy exactly similar equations, but with μ and ν as the independent variable in the respective cases.

As yet there are no restrictions on the values of H and K, so that whatever their values this equation possesses a general solution which may be written $E_n^K(\lambda)$, and it determines a solution of $\nabla^2 V = 0$, namely, the product

$$V = E_n^K(\lambda) E_n^K(\mu) E_n^K(\nu).$$

Since Laplace's equation is linear, it follows that any linear sum of solutions of this kind is also a solution.

Polynomial solutions of Laplace's equation

It will be shown that it is possible to choose the constants H and K so that Lamé's equation has solutions of the following forms:

    (i) a polynomial in λ;

or (ii) a polynomial in λ multiplied by *one* of the radical factors $\sqrt{(\lambda + a^2)}$, $\sqrt{(\lambda + b^2)}$, or $\sqrt{(\lambda + c^2)}$;

or (iii) a polynomial in λ multiplied by any *two* of these radical factors;

or (iv) a polynomial in λ multiplied by $\sqrt{[(\lambda + a^2)(\lambda + b^2)(\lambda + c^2)]}$.

Such a solution, if it exists, may be written in all cases, apart from a constant factor,

$$L(\lambda) = (\lambda + a^2)^{\kappa_1} (\lambda + b^2)^{\kappa_2} (\lambda + c^2)^{\kappa_3} (\lambda^m + a_1 \lambda^{m-1} + \ldots)$$
$$= (\lambda + a^2)^{\kappa_1} (\lambda + b^2)^{\kappa_2} (\lambda + c^2)^{\kappa_3} f(\lambda), \text{ say,}$$

where the indices κ_1, κ_2 and κ_3 are each independently equal to either 0 or $\frac{1}{2}$, m is a positive integer, and $f(\lambda)$ a polynomial of degree m.

If M and N denote respectively the corresponding expressions with μ and ν in place of λ, the product $L(\lambda) M(\mu) N(\nu)$ will be expressible, by means of (4) and (5), as a polynomial in x, y, z in which the terms of highest degree are of order $2(\kappa_1 + \kappa_2 + \kappa_3 + m)$. This polynomial will not of course be homogeneous. If its highest terms are of order n (its identity for integer cases with the n of (16) will appear shortly) we shall have always

$$\kappa_1 + \kappa_2 + \kappa_3 + m = \tfrac{1}{2} n,$$

so that if n is even all three or one of the κ's must be zero, and if n is odd two or none of the κ's must be zero.

If now this form for L is substituted in equation (15), it must become an identity, and this as a first condition requires H to be of a special form. To show this we have

$$A \frac{dL}{d\lambda} = \kappa_1 (\lambda + a^2)^{\kappa_1 - \frac{1}{2}} (\lambda + b^2)^{\kappa_2 + \frac{1}{2}} (\lambda + c^2)^{\kappa_3 + \frac{1}{2}} (\lambda^m + a_1 \lambda^{m-1} + \ldots)$$
$$+ \kappa_2 (\lambda + a^2)^{\kappa_1 + \frac{1}{2}} (\lambda + b^2)^{\kappa_2 - \frac{1}{2}} (\lambda + c^2)^{\kappa_3 + \frac{1}{2}} (\lambda^m + a_1 \lambda^{m-1} + \ldots)$$
$$+ \kappa_3 (\lambda + a^2)^{\kappa_1 + \frac{1}{2}} (\lambda + b^2)^{\kappa_2 + \frac{1}{2}} (\lambda + c^2)^{\kappa_3 - \frac{1}{2}} (\lambda^m + a_1 \lambda^{m-1} + \ldots)$$
$$+ m (\lambda + a^2)^{\kappa_1 + \frac{1}{2}} (\lambda + b^2)^{\kappa_2 + \frac{1}{2}} (\lambda + c^2)^{\kappa_3 + \frac{1}{2}} (\lambda^{m-1} + \ldots).$$

Considering at this stage only the highest power of λ, its order is

$$\kappa_1 + \kappa_2 + \kappa_3 + \tfrac{1}{2} + m = \tfrac{1}{2}(n+1)$$

and its coefficient is $\qquad \kappa_1 + \kappa_2 + \kappa_3 + m = \tfrac{1}{2}n.$

Hence when L is substituted in the left-hand side of (15), the term of highest order occurring will be

$$\tfrac{1}{4}n(n+1)\lambda^{\frac{1}{2}n+1}.$$

This must therefore equal the term of highest order on the right arising from $(H\lambda + K)L$, and accordingly we must have as a first condition for a solution possessing one of these special forms, and leading eventually to a harmonic polynomial of order n, $\qquad H = \tfrac{1}{4}n(n+1),$

where n is a positive integer.

Before proceeding to show that K also can be appropriately chosen for solutions of the present kind to exist, we notice that the polynomial solutions, in x, y, z, of $\nabla^2 V = 0$ can be classified under four different types. Table III shows the four cases and the relation between the degree of V the Lamé polynomial and L the Lamé function in the different cases. In this table F denotes a polynomial in x^2, y^2, z^2 of the form

$$F = \prod_{r=1}^{m}\left(\frac{x^2}{a^2+\delta_r} + \frac{y^2}{b^2+\delta_r} + \frac{z^2}{c^2+\delta_r} - 1\right).$$

TABLE III

Species	Form of solution in rectangular coordinates (Lamé polynomial of degree n)	Form of corresponding Lamé function L	Degree of $f(\lambda)$, the rational part of L	No. of values of K = no. of different L's
(i)	$V = F(x^2, y^2, z^2)$ n even	$f(\lambda)$	$\tfrac{1}{2}n$	$\tfrac{1}{2}(n+2)$
(ii)	$V = \begin{Bmatrix} x \\ y \\ z \end{Bmatrix} F(x^2,\, y^2,\, z^2)$ n odd	$\begin{Bmatrix} \sqrt{(\lambda+a^2)} \\ \sqrt{(\lambda+b^2)} \\ \sqrt{(\lambda+c^2)} \end{Bmatrix} f(\lambda)$	$\tfrac{1}{2}(n-1)$	$\tfrac{1}{2}(n+1)$
(iii)	$V = \begin{Bmatrix} yz \\ zx \\ xy \end{Bmatrix} F(x^2, y^2, z^2)$ n even	$\begin{Bmatrix} \sqrt{[(\lambda+b^2)(\lambda+c^2)]} \\ \sqrt{[(\lambda+c^2)(\lambda+a^2)]} \\ \sqrt{[(\lambda+a^2)(\lambda+b^2)]} \end{Bmatrix} f(\lambda)$	$\tfrac{1}{2}(n-2)$	$\tfrac{1}{2}n$
(iv)	$V = xyz F(x^2, y^2, z^2)$ n odd	$\sqrt{[(\lambda+a^2)(\lambda+b^2)(\lambda+c^2)]}\, f(\lambda)$	$\tfrac{1}{2}(n-3)$	$\tfrac{1}{2}(n-1)$

The last column in the table gives the number of different values of K that can be assigned, for given n, in order that the solution shall be of the form shown in the third column. We proceed now to demonstrate that K can be suitably chosen. There are four cases corresponding to the four species of Table III.

Determination of K

(I) For functions of species (i) the value of n must be even, and the function $L(\lambda)$ will then be a polynomial of degree $m = \frac{1}{2}n$. It therefore contains $\frac{1}{2}n + 1$ coefficients of which that of the highest power has already been taken as unity. Thus we have

$$L = \lambda^{\frac{1}{2}n} + a_1 \lambda^{\frac{1}{2}n-1} + a_2 \lambda^{\frac{1}{2}n-2} + \dots,$$

where the coefficients $a_1, a_2, \dots$ are to be disposed in such a way that this is a solution of Lamé's equation. For this, the substitution of this expression in the differential equation for L

$$A^2 \frac{d^2 L}{d\lambda^2} + A \frac{dA}{d\lambda} \cdot \frac{dL}{d\lambda} = (H\lambda + K)L$$

must produce an identity. As already seen, the highest power of λ occurring on the left is $\frac{1}{2}n + 1$, and the resulting expression must contain $\frac{1}{2}n + 2$ coefficients corresponding to the terms in $\lambda^{\frac{1}{2}n+1}, \lambda^{\frac{1}{2}n}, \dots, \lambda, \lambda^0$. Because of the choice of $H = \frac{1}{4}n(n+1)$ the terms of highest degree are the same on both sides, and there will remain $\frac{1}{2}n + 1$ coefficients that must be equal on the two sides. The expression of this yields $\frac{1}{2}n + 1$ equations, and these are evidently linear in K and in the coefficients, $\frac{1}{2}n + 1$ in number, of the polynomial assumed for L. Elimination of these coefficients gives an equation for K that is essentially of degree $\frac{1}{2}n + 1$, and accordingly there are $\frac{1}{2}n + 1$ possible values of K to each of which corresponds a polynomial solution. This is the number given in the fourth column of the table on p. 57.

It will be shown later that all the roots of this equation for K are real and different and lead in all to $\frac{1}{2}n + 1$ different polynomial solutions.

(II) Consider next functions of species (iii), so that again n is even.

In this case L will contain two radical factors (which can be selected in three ways) so that the degree of the rational part of $L(\lambda)$ is now $\frac{1}{2}n - 1$. It accordingly contains $\frac{1}{2}n$ coefficients, and by the same procedure as in the previous case their elimination leads to $\frac{1}{2}n$ values of K. Thus with each of the three possible pairs of radical factors there are associated $\frac{1}{2}n$ values of K.

Again, it will be shown later that these values of K are all real and different and lead to different L functions.

If n is even only functions of the species (i) and (iii) exist and the total number of Lamé functions of given even order n is accordingly

$$(\tfrac{1}{2}n + 1) + 3(\tfrac{1}{2}n) = 2n + 1.$$

(III) Functions of species (ii) exist when n is odd. The one radical factor can now be chosen in three ways, and the rational part of L will be of degree $\frac{1}{2}(n-1)$. It will contain $\frac{1}{2}(n+1)$ coefficients, and their elimination leads to $\frac{1}{2}(n+1)$ values of K, which will be shown to be all real and different.

(IV) Functions of species (iv) also exist when n is odd. $L(\lambda)$ now contains all three radical factors, and its rational part is of degree $\frac{1}{2}(n-3)$. There are accordingly $\frac{1}{2}(n-1)$ different values of K, which can be shown to be all real and different, leading to $\frac{1}{2}(n-1)$ different Lamé functions.

If n is odd the Lamé functions are of the species (ii) and (iv) only and their total number is
$$3 \cdot \tfrac{1}{2}(n+1) + \tfrac{1}{2}(n-1) = 2n+1.$$

Thus, whether n is odd or even, the total number of Lamé functions is $2n+1$. This result is exactly what would be expected, for each of the Lamé functions $L(\lambda)$ leads to a Lamé polynomial $V(x,y,z)$ satisfying $\nabla^2 V = 0$, and the terms of highest degree in V give rise to an associated homogeneous polynomial, V_h say, of order n also satisfying $\nabla^2 V_h = 0$. Now it may easily be proved by counting coefficients that there are, quite generally, $2n+1$ independent homogeneous polynomial solutions of degree n. The terms of highest degree in each of the Lamé polynomials will evidently be a certain linear sum of such harmonics specially adapted to the ellipsoidal coordinates.

The terms of lower orders in the Lamé polynomial $V(x,y,z)$ are determined uniquely by the terms of highest order, for we have

$$V = \left\{ 1 \quad \begin{matrix} x & yz \\ y & zx \\ z & xy \end{matrix} \quad xyz \right\} \prod_1^m \left\{ \frac{x^2}{a^2+\delta_r} + \frac{y^2}{b^2+\delta_r} + \frac{z^2}{c^2+\delta_r} - 1 \right\},$$

and the homogeneous part given by the terms of highest degree is

$$V_h = \left\{ 1 \quad \begin{matrix} x & yz \\ y & zx \\ z & xy \end{matrix} \quad xyz \right\} \prod_1^m \left\{ \frac{x^2}{a^2+\delta_r} + \frac{y^2}{b^2+\delta_r} + \frac{z^2}{c^2+\delta_r} \right\},$$

so that if V_h is given the expression for V can be obtained at once.

It will be shown later that the $2n+1$ homogeneous polynomials arising from the Lamé polynomials of degree n are themselves linearly independent, and it will then follow that any homogeneous harmonic polynomial of degree n is uniquely expressible as a linear sum of them.

The Lamé functions and polynomials of order 0, 1, 2, and 3

By straightforward application of the foregoing theory it is readily found that these are as follows:

The 1 Lamé function of order 0 is 1 (or any constant), corresponding to the solution $V = \text{constant}$ of Laplace's equation.

The 3 Lamé functions of order 1 are
$$\sqrt{(\lambda+a^2)}, \quad \sqrt{(\lambda+b^2)}, \quad \sqrt{(\lambda+c^2)},$$
and these correspond to the harmonic functions x, y, and z.

The 5 Lamé functions of order 2 are
$$\sqrt{[(\lambda+b^2)(\lambda+c^2)]}, \quad \sqrt{[(\lambda+c^2)(\lambda+a^2)]}, \quad \sqrt{[(\lambda+a^2)(\lambda+b^2)]},$$
together with the two functions
$$\lambda + \tfrac{1}{3}(a^2+b^2+c^2) \pm \tfrac{1}{3}\{a^4+b^4+c^4-b^2c^2-c^2a^2-a^2b^2\}^{\frac{1}{2}}.$$

The first three correspond respectively to the second-order harmonics yz, zx, and xy, and the last two have for their homogeneous parts functions of the form $j(x^2-y^2)+k(x^2-z^2)$, where j and k are certain constants, of which there are two independent types.

The 7 Lamé functions of order 3 are $\sqrt{[(\lambda+a^2)(\lambda+b^2)(\lambda+c^2)]}$ and the six functions obtained by cyclic interchange of a, b, and c in

$$\sqrt{(\lambda+a^2)}\{\lambda+\tfrac{1}{5}(a^2+2b^2+2c^2)\pm\tfrac{1}{5}(a^4+4b^4+4c^4-7b^2c^2-c^2a^2-a^2b^2)^{\frac{1}{2}}\}.$$

These correspond to xyz together with six polynomials containing terms of degree 3 and lower whose homogeneous parts constitute six independent linear sums of x^3-3xy^2, x^3-3xz^2, y^3-3yz^2, y^3-3x^2y, z^3-3x^2z, and z^3-3y^2z.

The relation of Lamé polynomials to spherical harmonics

Before proceeding to prove that the K's of Lamé's equation are all real and different it is necessary to consider the analogy between the Lamé polynomials, which are adapted for use in reference to ellipsoids, and the ordinary spherical harmonics. The relationship is of special importance in the case of spheroids, when $a=b$ or $b=c$, for then the Lamé functions are expressible in terms of certain well-known functions, as will be seen later. The results to be obtained, in some instances however, involve a limiting process, and cannot be arrived at simply by setting $a=b$ in the general Lamé functions.

Suppose variables θ,ϕ are introduced replacing μ,ν and defined by

$$\left.\begin{aligned} x &= \sqrt{(\lambda+a^2)}\sin\theta\cos\phi, \\ y &= \sqrt{(\lambda+b^2)}\sin\theta\sin\phi, \\ z &= \sqrt{(\lambda+c^2)}\cos\theta. \end{aligned}\right\} \tag{17}$$

Equations (5) show that θ,ϕ depend on μ,ν only. It is important to notice that these variables are *not* the θ,ϕ of ordinary spherical polar coordinates. They do, however, tend to them as $\lambda\to\infty$, for then the radical factors in x,y,z in (17) all tend to equality with r. It has already been seen that μ and ν are always finite, and it is clear also that for $\lambda=\infty$ they depend only on the ratios $x:y:z$.

A Lamé polynomial corresponding to a certain normal solution $L(\lambda)M(\mu)N(\nu)$ will at great distance from the origin be effectively equal to $\lambda^{\frac{1}{2}n}M(\mu)N(\nu)$. But since $\lambda\to r^2$ this will also be everywhere exactly the value of the homogeneous part of the Lamé polynomial. But this is a spherical harmonic, and since the coordinates θ,ϕ tend to the θ,ϕ of spherical polars we must have accurately

$$MN = F(\sin\theta\cos\phi,\ \sin\theta\sin\phi,\ \cos\theta)$$

where F is a spherical surface harmonic function of θ,ϕ. Since this is an identity it must hold everywhere, but it is only at infinity that θ,ϕ become the ordinary polar angles; elsewhere they are defined by (17) and have no simple geometrical meanings. Thus when expressed in terms of θ,ϕ so defined, the product MN always takes the form of a spherical surface harmonic. But it is a known property that any continuous function of θ, ϕ can be expressed as a linear sum of independent surface harmonics. It follows that any function of μ, ν can be expressed as a linear sum of the form

$$\Sigma A_k MN.$$

The sum will in general require harmonics of all orders, so that there will be $2n+1$ terms involving the $2n+1$ harmonics of order n; that is, 1 constant term, 3 terms of first order, and so on.

Oblate spheroids $a = b$

To consider the form the functions take when the confocal quadrics are surfaces of revolution round the z-axis, let it be supposed that $b \to a$. Since

$$\lambda > -c^2 > \mu > -b^2 > \nu > -a^2,$$

it is clear that ν can no longer serve as a parameter when $a = b$, for it is always equal to $-a^2$. A third coordinate must therefore be introduced to replace ν. For this purpose we have from (5)

$$\frac{y^2}{x^2} = -\frac{(b^2+\lambda)(b^2+\mu)(b^2+\nu)}{(a^2+\lambda)(a^2+\mu)(a^2+\nu)} \cdot \frac{(a^2-b^2)(a^2-c^2)}{(a^2-b^2)(b^2-c^2)}$$

$$= -\frac{b^2+\nu}{a^2+\nu} \quad \text{when} \quad a = b$$

since the factors (a^2-b^2) cancel accurately always, while pairs of factors such as $b^2+\lambda$ and $a^2+\lambda$ become equal and non-zero and then also cancel. It is therefore necessary to retain only $b^2+\nu$ and $a^2+\nu$. Hence we have

$$-\frac{b^2+\nu}{a^2+\nu} = \tan^2\phi,$$

where ϕ is in this case everywhere the ordinary azimuth angle of spherical polar coordinates. If the parameter ν is replaced by ϕ, so that planes through the z-axis are now one of the three orthogonal families, we have

$$\nu = -a^2\sin^2\phi - b^2\cos^2\phi,$$

and hence $\quad \cos\phi = \underset{b \to a}{\text{Lt}} \sqrt{\left(\frac{\nu+a^2}{a^2-b^2}\right)}, \quad \sin\phi = \underset{b \to a}{\text{Lt}} \sqrt{\left(-\frac{\nu+b^2}{a^2-b^2}\right)},$

and we notice that ϕ depends only on ν, and the function of ϕ replacing N in the normal solution is to be obtained by a limiting process involving the foregoing relations between ϕ and ν.

Where θ is concerned, μ can take values in the range $-c^2$ to $-b^2$, and no limiting process is required. We can thus put $a = b$ and make x, y, z large to find the relation between them. So that

$$\frac{x^2+y^2}{a^2+\mu} + \frac{z^2}{c^2+\mu} = 1$$

becomes

$$\frac{\sin^2\theta}{a^2+\mu} + \frac{\cos^2\theta}{c^2+\mu} = 0.$$

Hence
$$\sin \theta = \sqrt{\left(\frac{a^2+\mu}{a^2-c^2}\right)} \quad \text{and} \quad \cos \theta = \sqrt{\left(-\frac{c^2+\mu}{a^2-c^2}\right)}.$$

Thus when $a = b$ the parameter μ depends on θ only. It is to be remembered that θ is defined by (17) and reduces to the spherical polar angle only at great distance.

Since, however, the product MN is quite generally a spherical surface harmonic in θ and ϕ as here defined, for spheroids it must become such that M is a function of θ only, and N a function of ϕ only. But when ordinary spherical surface harmonics are put in normal form, the functions depending on θ only are the Legendre functions $P_n^p(\cos \theta)$, and those depending on ϕ only are the circular functions $\frac{\cos}{\sin} p\phi$. Hence we must have

$$\left. \begin{aligned} M(\mu) &= P_n^p(\cos \theta), \\ N(\nu) &= \frac{\cos}{\sin} p\phi \end{aligned} \right\} \quad (p = 0, 1, \ldots, \text{ or } n). \tag{18}$$

Since no limiting process is involved in the relation between μ and θ, the form of $M(\mu)$ arrived at here must be simply the result of writing $a = b$ in the general form of M. Also, the function $L(\lambda)$ in the case $a = b$ can be got directly from $M(\mu)$ simply by writing λ for μ. On the other hand it cannot be obtained from the present form for N since this has resulted from a limiting process.

We have, in fact, apart from an arbitrary constant factor which may be omitted

$$M = P_n^p(t) \quad \text{where} \quad t = \cos \theta$$

$$= (1-t^2)^{\frac{1}{2}p} D^{p+n}(1-t^2)^n$$

where $D = \dfrac{d}{dt}$, and p can take any of the values $0, 1, 2, \ldots, n$. If we write

$$t^2 = \cos^2 \theta = -\frac{c^2+\mu}{a^2-c^2} = -\tau^2, \quad \text{say},$$

then
$$M = (1+\tau^2)^{\frac{1}{2}p} D^{p+n}(1+\tau^2)^n \tag{19}$$

and τ, so defined, is imaginary for the function M since $-c^2-\mu > 0$.

To find the function $L(\lambda)$ we now simply replace μ by λ in the foregoing, so that
$$L = (1+\tau^2)^{\frac{1}{2}p} D^{p+n}(1+\tau^2)^n, \tag{20}$$

where now
$$\tau^2 = \frac{c^2+\lambda}{a^2-c^2}, \tag{21}$$

and τ is now a real variable since $\lambda+c^2 > 0$.

To each value of $p = 1, 2, \ldots$ or n, there correspond two harmonic functions, LMN, because of the alternative N factor $\cos p\phi$ or $\sin p\phi$. But for $p = 0$, there is only one N factor, namely 1. We thus arrive at $2n+1$ harmonics in all, as would be expected.

Prolate spheroids $b = c$

If $b \to c$, it is clear, since $-c^2 > \mu > -b^2$, that in this case μ can no longer serve as parameter. To deal with this feature we define θ' and ψ by means of

$$x = \sqrt{(\lambda + a^2)} \cos \theta',$$
$$y = \sqrt{(\lambda + b^2)} \sin \theta' \cos \psi,$$
$$z = \sqrt{(\lambda + c^2)} \sin \theta' \sin \psi,$$

so that for $b = c$ the coordinate ψ is the azimuth angle in the yz-plane, but only at great distance does θ' become the polar angle (measured from the x-axis). To find the relation between μ and ψ, we have (in a similar way to the case $a = b$)

$$\frac{y^2}{z^2} = -\frac{b^2 + \mu}{c^2 + \mu} = \cot^2 \psi,$$

so that $$\mu = -b^2 \sin^2 \psi - c^2 \cos^2 \psi.$$

Hence $$\cos \psi = \operatorname*{Lt}_{b \to c} \sqrt{\left(\frac{\mu + b^2}{b^2 - c^2} \right)} \quad \text{and} \quad \sin \psi = \operatorname*{Lt}_{b \to c} \sqrt{\left(-\frac{\mu + c^2}{b^2 - c^2} \right)},$$

and the function M will depend only on ψ.

To find the relation between θ' and ν we put $b = c$ and allow x, y, z to become large, whereupon the equation to the surfaces

$$\frac{x^2}{a^2 + \nu} + \frac{y^2 + z^2}{c^2 + \nu} = 1$$

becomes at great distance $$\frac{\cos^2 \theta'}{a^2 + \nu} + \frac{\sin^2 \theta'}{c^2 + \nu} = 0$$

so that $$\sin \theta' = \sqrt{\left(-\frac{c^2 + \nu}{a^2 - c^2} \right)}, \quad \cos \theta' = \sqrt{\left(\frac{a^2 + \nu}{a^2 - c^2} \right)},$$

and no limiting process is involved in these relations. The function N now depends only on θ'.

In the same way as before, since MN when expressed in terms of θ', ψ must have the form of a spherical surface harmonic, we must now have

$$\left. \begin{aligned} M &= {\textstyle\genfrac{}{}{0pt}{}{\cos}{\sin}} p\psi, \\ N &= P_n^p(\cos \theta') \end{aligned} \right\} \quad (p = 0, 1, \ldots, n). \tag{22}$$

In the present case N is simply the result of putting $b = c$ in the general Lamé function, whereas M is obtained by a limiting process. We have, in fact,

$$N = (1 - t^2)^{\frac{1}{2}p} D^{p+n} (1 - t^2)^n, \tag{23}$$

where $$t^2 = \cos^2 \theta' = \frac{a^2 + \nu}{a^2 - c^2}.$$

Hence to find $L(\lambda)$, we write $$t^2 = \frac{a^2 + \lambda}{a^2 - c^2}, \tag{24}$$

and then since $1 - t^2$ is now negative we always obtain L in real form by writing

$$L = (t^2 - 1)^{\frac{1}{2}p} D^{p+n} (t^2 - 1)^n. \tag{25}$$

The Lamé functions that reduce when $a = b$ to $\cos p\phi$ and $\sin p\phi$.

To examine the question which of the general Lamé functions reduce to $\cos p\phi$ and which to $\sin p\phi$, it has been seen that $M(\mu)N(\nu)$ when expressed in terms of θ, ϕ takes the form of a tesseral harmonic, thus

$$M(\mu)N(\nu) = P_n^p(\cos\theta) \, {\cos \atop \sin} p\phi \quad (p = 0, 1, ..., n).$$

If on the right-hand side of this relation we wish to replace θ and ϕ by μ and ν, then, apart from constant factors irrelevant to the present argument, the appropriate relations are

$$(x) \qquad \sin\theta\cos\phi \propto \sqrt{[(a^2+\mu)(a^2+\nu)]},$$

$$(y) \qquad \sin\theta\sin\phi \propto \sqrt{[(b^2+\mu)(b^2+\nu)]},$$

$$(z) \qquad \cos\theta \propto \sqrt{[(c^2+\mu)(c^2+\nu)]}.$$

Also
$$P_n^p(\cos\theta) \propto (\sin\theta)^p \left(\frac{d}{dt}\right)^p P_n(t) \quad \text{where} \quad t = \cos\theta,$$

and $P_n(t)$ is the ordinary Legendre polynomial of order n. We thus have

$$P_n^p(\cos\theta)\{\cos p\phi + i\sin p\phi\} = \left\{\left(\frac{d}{dt}\right)^p P_n(t)\right\} \{\sin\theta(\cos\phi + i\sin\phi)\}^p$$

$$= (\text{polynomial in } \cos\theta) \times (x+iy)^p,$$

omitting constant multiplying factors that make no difference to the particular points we wish to establish.

(a) Suppose now that p is even

The solution involving $\cos p\phi$ is the real part of the left-hand side in this equation, and hence requires on the right-hand side the real part of $(x+iy)^p$. But we have for this (where the coefficients c_r denote constants),

$$\Re(x+iy)^p = x^p - c_1 x^{p-2}y^2 + c_2 x^{p-4}y^4 - ...$$

$$= \text{rational polynomial in } x^2 \text{ and } y^2.$$

Hence a solution of the form $V(x^2, y^2, z^2)$, for the case $a = b$ will correspond to an even value of p and the term $\cos p\phi$ for the N function. (See line 1 of Table IV.)

On the other hand, $\sin p\phi$ evidently requires the imaginary part of $(x+iy)^p$, and we have

$$\Im(x+iy)^p = px^{p-1}y - ... \pm pxy^{p-1}$$

$$= xy \, (\text{rational polynomial in } x^2 \text{ and } y^2).$$

Hence a solution originally of the form $xyF(x^2, y^2, z^2)$ will correspond to p even and the function $\sin p\phi$ for N. (This is shown in line 7 of Table IV.)

(b) Suppose that p is odd

A solution involving $\cos p\phi$ again requires the real part of the right-hand side, but now

$$\Re(x+iy)^p = x(x^{p-1} - c_1 x^{p-3}y^2 + ...)$$

$$= x \, (\text{rational polynomial in } x^2 \text{ and } y^2).$$

Hence a solution of the form $xF(x^2, y^2, z^2)$ will correspond to an odd value of p and $\cos p\phi$ for N. (Line 2 of Table IV.)

Similarly

$$\Im(x+iy)^p = y \text{ (rational polynomial in } x^2 \text{ and } y^2)$$

and hence a solution of the form $yF(x^2, y^2, z^2)$ requires p odd and $\sin p\phi$ for the N function.

Finally, since z depends only on θ, the presence or absence of z as a factor of the polynomial solution will not affect the foregoing results. That is, a solution of the form zF has the same (odd or even) value of p and function for N as does a solution of the form F. We can thus set out in a table the form of the N-function and the appropriate p (odd or even) for the eight possible forms of ellipsoidal harmonics for the case $a = b$.

TABLE IV

Form of general Lamé polynomial	Form of general L: $f(\lambda)$ rational	$a = b$	
		p	N
$F(x^2, y^2, z^2)$	$f(\lambda)$	even	$\cos p\phi$
xF	$\sqrt{(\lambda+a^2)}f(\lambda)$	odd	$\cos p\phi$
yF	$\sqrt{(\lambda+b^2)}f(\lambda)$	odd	$\sin p\phi$
zF	$\sqrt{(\lambda+c^2)}f(\lambda)$	even	$\cos p\phi$
yzF	$\sqrt{[(\lambda+b^2)(\lambda+c^2)]}f(\lambda)$	odd	$\sin p\phi$
zxF	$\sqrt{[(\lambda+c^2)(\lambda+a^2)]}f(\lambda)$	odd	$\cos p\phi$
xyF	$\sqrt{[(\lambda+a^2)(\lambda+b^2)]}f(\lambda)$	even	$\sin p\phi$
$xyzF$	$\sqrt{[(\lambda+a^2)(\lambda+b^2)(\lambda+c^2)]}f(\lambda)$	even	$\sin p\phi$

Harmonics of orders 1 and 2 for the case $a = b$

When $a = b$, we have (apart from constant factors)

$$x \propto \sqrt{[(a^2+\lambda)(a^2+\mu)]} \cdot \sqrt{\left(\frac{a^2+\nu}{a^2-b^2}\right)} = \sqrt{[(a^2+\lambda)(a^2+\mu)]}\cos\phi,$$

$$y \propto \sqrt{[(a^2+\lambda)(a^2+\mu)]} \cdot \sqrt{\left(\frac{b^2+\nu}{a^2-b^2}\right)} = \sqrt{[(a^2+\lambda)(a^2+\mu)]}\sin\phi,$$

$$z \propto \sqrt{[(c^2+\lambda)(c^2+\mu)]},$$

since $\nu = -a^2$ and no limiting process is involved.

As illustrations of the foregoing theory we next evaluate the harmonics of order 1 and 2 when $a = b$.

(i) *The three harmonics of order* 1

Here $n = 1$, and hence we have

$$L = (1+\tau^2)^{\frac{1}{2}p} D^{p+1}(1+\tau^2) \quad (p = 0, 1),$$

where
$$\tau^2 = (c^2+\lambda)/(a^2-c^2).$$

If $p = 0$:
$$L = D(1+\tau^2)$$
$$= \tau = \sqrt{(c^2+\lambda)} \quad \text{omitting constant factors.}$$

The function LMN then reduces to $\sqrt{[(c^2+\lambda)(c^2+\mu)]}$, which corresponds to the ellipsoidal harmonic z.

If p = 1:
$$L = (1+\tau^2)^{\frac{1}{2}} D^2(1+\tau^2)$$
$$= (1+\tau^2)^{\frac{1}{2}} = \sqrt{(a^2+\lambda)}.$$

The function LMN can now take either of the forms

$$\sqrt{[(a^2+\lambda)(a^2+\mu)]} \; {\textstyle{\cos \atop \sin}} \; \phi.$$

By the preceding table, line 2 shows that the first of these corresponds to the harmonic x, and line 3 shows that the second corresponds to y.

(ii) *The five harmonics of order 2*

Here $n = 2$ and we have for the general form of L

$$L = (1+\tau^2)^{\frac{1}{2}p} D^{p+2}(1+\tau^2)^2 \quad (p = 0, 1, 2).$$

If p = 0: $L = D^2(1+\tau^2)^2$
$$= 1+3\tau^2 = a^2+2c^2+3\lambda \quad \text{apart from constant factors.}$$

Hence the solution LMN is $(3\lambda+a^2+2c^2)(3\mu+a^2+2c^2)$.

The corresponding harmonic in x, y, z is easily found, since λ and μ are the roots of
$$\frac{x^2+y^2}{a^2+\lambda}+\frac{z^2}{c^2+\lambda} = 1,$$

so that
$$\lambda\mu = a^2c^2 - c^2(x^2+y^2) - a^2z^2,$$
$$\lambda+\mu = x^2+y^2+z^2-a^2-c^2.$$

In rectangular coordinates the solution is therefore

$$3(a^2+2c^2)(x^2+y^2+z^2) - 9c^2(x^2+y^2) - 9a^2z^2 + 9a^2c^2 - 3(a^2+2c^2)(a^2+c^2) + (a^2+2c^2)^2.$$

The constant portion of this can be omitted and the rest is easily seen to be of the form $j(x^2-y^2)+k(x^2-z^2)$, where j and k are constants.

If p = 1: $L = (1+\tau^2)^{\frac{1}{2}} D^3(1+\tau^2)^2 = \tau\sqrt{(1+\tau^2)} = \sqrt{[(a^2+\lambda)(c^2+\lambda)]}.$

The LMN solutions are therefore $\sqrt{[(a^2+\lambda)(c^2+\lambda)(a^2+\mu)(c^2+\mu)]} \; {\textstyle{\cos \atop \sin}} \; \phi.$ Line 6 of Table IV shows that the solution with factor $\cos\phi$ gives the second-order harmonic xz, and line 5 shows that the one with $\sin\phi$ as factor gives yz.

If p = 2: $L = (1+\tau^2) D^4(1+\tau^2)^2 = 1+\tau^2 = a^2+\lambda.$

The LMN solutions are therefore $(a^2+\lambda)(a^2+\mu) \; {\textstyle{\cos \atop \sin}} \; 2\phi$, which correspond respectively to x^2-y^2 (by line 1 of Table IV) and xy (by line 7).

These results may be compared with those already given on page 59. The solutions when expressed in terms of x, y, z must obviously reduce to ordinary spherical harmonics, but the importance of the present analysis is that it enables us to find the particular linear sums of these appropriate to an ellipsoid in the general case, and to certain spheroids in the cases $a = b$, and $b = c$.

The values of K are all real and different

Having established the form of the Lamé functions in the case of spheroids, we are now in a position to return to the question of the values that K must have in the general case $(a > b > c)$ in order to give a solution of Lamé's equation in the form of a polynomial multiplied by 0, 1, 2 or 3 of the radical factors.

Supposing first that $a = b$, we have seen that ν must be replaced by ϕ related to it by

$$\nu = -a^2 \sin^2 \phi - b^2 \cos^2 \phi.$$

The equation for the function N, in its general form, is

$$C \frac{d}{d\nu}\left(C \frac{dN}{d\nu}\right) = (H\nu + K)N.$$

Now
$$C^2 = (a^2 + \nu)(b^2 + \nu)(c^2 + \nu) = -(a^2 - b^2)^2 \sin^2 \phi \cos^2 \phi (c^2 + \nu),$$

so that
$$C = i(a^2 - b^2)\sin\phi\cos\phi\sqrt{(c^2 + \nu)}.$$

Also
$$d\nu = -2(a^2 - b^2)\sin\phi\cos\phi\, d\phi$$

and hence
$$C\frac{d}{d\nu} = -\tfrac{1}{2}i\sqrt{(c^2 + \nu)}\frac{d}{d\phi}$$

and the equation for N when ϕ is the independent variable is

$$-(c^2 + \nu)\frac{d^2 N}{d\phi^2} - \frac{1}{2}\frac{d\nu}{d\phi}\cdot\frac{dN}{d\phi} = 4(H\nu + K)N.$$

If now we let $b \to a$, then in the limit $\nu = -a^2$ and $\frac{d\nu}{d\phi} = 0$, and the equation for N reduces to

$$(a^2 - c^2)\frac{d^2 N}{d\phi^2} + \{n(n+1)a^2 + 4K\}N = 0.$$

From the discussion of the preceding sections it is necessary that this shall be satisfied in turn by

$$N = \genfrac{}{}{0pt}{}{\cos}{\sin}p\phi \qquad (p = 0, 1, 2, \ldots, n).$$

Accordingly we must have

$$n(n+1)a^2 + 4K = p^2(a^2 - c^2)$$

or
$$4K = p^2(a^2 - c^2) - n(n+1)a^2.$$

It is thus seen that for $a = b$, the equation giving the values of K must have these $n+1$ roots corresponding to $p = 0, 1, 2, \ldots, n$, and they are obviously all real and different, and the roots corresponding to odd values of p separate those corresponding to even values.

It may now readily be proved that in the general case when $a \neq b$ the preceding result still holds in the form that for Lamé functions of any one type of given order n the values of K are all real and different and separate those corresponding to any other type of the same order.

To show this, let us consider first the case when n is even. When $a = b$ it has been seen there are $n+1$ values of K. From Table IV it is seen that for p even $(0, 2, \ldots)$ the general solutions $(a \neq b)$ are of the form

$$f(\lambda) \quad \text{and} \quad \sqrt{[(\lambda + a^2)(\lambda + b^2)]} f(\lambda),$$

while for p odd $(1, 3, \ldots)$ the general solutions are of the form

$$\sqrt{[(\lambda + b^2)(\lambda + c^2)]} f(\lambda) \quad \text{and} \quad \sqrt{[(\lambda + c^2)(\lambda + a^2)]} f(\lambda),$$

where as usual $f(\lambda)$ denotes a polynomial in λ. Also, in the case $a = b$, the roots in K for even values of p separate those for odd values. Supposing now that b is allowed to vary slowly away from a so that $a \neq b$, the foregoing property could cease to hold only if at some stage a root K corresponding to a function of the first group

$$f(\lambda) \quad \text{or} \quad \sqrt{[(\lambda + a^2)(\lambda + b^2)]} f(\lambda)$$

became equal to a root K corresponding to a function of the second group

$$\sqrt{[(\lambda + b^2)(\lambda + c^2)]} f(\lambda) \quad \text{or} \quad \sqrt{[(\lambda + c^2)(\lambda + a^2)]} f(\lambda).$$

It will now be shown that this cannot possibly occur. For if it did, let the two corresponding L-functions having the same values of K be L_1 and L_2. Then if u is defined by

$$d\lambda = -2\sqrt{[(\lambda + a^2)(\lambda + b^2)(\lambda + c^2)]}\, du,$$

Lamé's equation for them may be written

$$\frac{d^2 L_1}{du^2} + (H\lambda + K)L_1 = 0,$$

$$\frac{d^2 L_2}{du^2} + (H\lambda + K)L_2 = 0.$$

Hence in the customary way, dashes denoting differentiation with respect to u,

$$L_1'' L_2 - L_2'' L_1 = 0,$$

which integrates to

$$L_1' L_2 - L_2' L_1 = \text{constant} = C, \text{ say,}$$

or

$$\frac{d}{du}(L_1/L_2) = C/L_2^2,$$

which in terms of λ becomes

$$-2\sqrt{[(\lambda + a^2)(\lambda + b^2)(\lambda + c^2)]} \frac{d}{d\lambda}(L_1/L_2) = C/L_2^2. \tag{26}$$

It will be shown that unless $C = 0$ this equation cannot be satisfied.

First it is seen that whatever the form of L_2 its square, L_2^2, is rational. Now a possible form for L_1/L_2, with an obvious notation, is

$$f_1(\lambda)/\sqrt{[(\lambda + b^2)(\lambda + c^2)]} f_2(\lambda),$$

and differentiation of this with respect to λ cannot introduce a radical factor $\sqrt{(\lambda + a^2)}$ anywhere. But the left-hand side of (26) contains $\sqrt{(\lambda + a^2)}$, and hence the equation cannot hold unless $C = 0$, in which case it reduces to

$$\frac{d}{d\lambda}(L_1/L_2) = 0, \quad \text{or} \quad L_1 = kL_2.$$

An exactly similar argument holds in each of the remaining possibilities for L_1/L_2. But the two L-functions are essentially of different species and hence cannot bear a constant ratio to each other. Hence the initial assumption is false and their K values cannot be equal; therefore the K values corresponding to one species must always separate those of any other species but of the same order. It follows also that only real values of K are possible, since, as b changes, two real values would have to become equal before complex or imaginary roots could come into existence.

The whole of the preceding argument applies in exactly the same way for n odd but with the species of functions now changed in accordance with the forms indicated by Table IV.

We may therefore conclude that for functions of a given species and order, the general equation for K when $a > b > c$ has all its roots real and different.

The linear independence of the $2n+1$ Lamé functions of a given order

The Lamé functions of order n satisfy

$$A^2 \frac{d^2 L}{d\lambda^2} + A \frac{dA}{d\lambda} \frac{dL}{d\lambda} - \tfrac{1}{4} n(n+1) L = KL,$$

and denoting by Δ the whole of the operation affecting L on the left-hand side, this equation can be written shortly*

$$\Delta L = KL.$$

Supposing now that a linear relation holds between the $2n+1$ possible Lamé functions, of the form

$$\sum_1^{2n+1} c_s L_s \equiv 0 \quad (s = 1, 2, \ldots, 2n+1),$$

wherein the c_s are constants not all zero, it is plain that if we break up the relation into parts in each of which the radical factors are the same, then the several linear sums so obtained must each vanish identically. For example, if n were even, there would be $\tfrac{1}{2}n+1$ rational Lamé functions (of the first species) and three sets each of $\tfrac{1}{2}n$ Lamé functions containing a pair of radical factors. Evidently in an identically zero linear sum of these $2n+1$ functions, the rational part, the part containing $\sqrt{[(\lambda+b^2)(\lambda+c^2)]}$, the part containing $\sqrt{[(\lambda+c^2)(\lambda+a^2)]}$, and the part containing $\sqrt{[(\lambda+a^2)(\lambda+b^2)]}$ must all vanish separately. The same must hold when n is odd, and the parts containing $\sqrt{(\lambda+a^2)}$, $\sqrt{(\lambda+b^2)}$, $\sqrt{(\lambda+c^2)}$, and $\sqrt{[(\lambda+a^2)(\lambda+b^2)(\lambda+c^2)]}$ must vanish separately.

If we select any one of these parts it will be a linear sum

$$\sum_{s=1}^{j} c_s L_s \equiv 0 \tag{27}$$

in which the number of L functions, j, is equal to the number of different possible values of K, as given in the last column of Table III. If $K_1, K_2, \ldots, K_j$ denote

---

* There will be no confusion with the earlier use of Δ in Chapter IV.

these different values of K we shall have for each L_s

$$\Delta L_s = K_s L_s \quad (s = 1, 2, ..., j).$$

Hence, operating on (27) with Δ we get

$$\sum_{s=1}^{j} K_s c_s L_s = 0.$$

A second operation with Δ gives

$$\sum_{s=1}^{j} K_s^2 c_s L_s = 0,$$

and so on. Operating successively $j-1$ times, we obtain in all j linear relations, and if the j quantities $c_s L_s$ are eliminated we obtain

$$\begin{vmatrix} 1 & 1 & 1 & \cdots & \cdots & 1 \\ K_1 & K_2 & K_3 & \cdots & \cdots & K_j \\ \cdots & \cdots & \cdots & \cdots & \cdots & \cdots \\ K_1^{j-1} & K_2^{j-1} & K_3^{j-1} & \cdots & \cdots & K_j^{j-1} \end{vmatrix} = 0.$$

But this equation is exactly equivalent to

$$\prod_{1}^{j} (K_p - K_q) = 0 \quad (p \neq q),$$

and such a result cannot be true, since all the K_s are different. Accordingly there cannot subsist a linear relation between the Lamé functions of a given order, and their linear independence is thus established.

The zeros of the Lamé functions

In the present section we will denote by L_r the rational part of any selected Lamé function, that is, when radical factors, if any, are omitted. Radical factors, if present in a Lamé function, imply zeros at $\lambda = -a^2$, $\lambda = -b^2$, or $\lambda = -c^2$, so that if L is regarded as a polynomial in $\lambda + a^2$, as is found convenient in the sequel, the presence of radical factors will mean zeros of L at $\lambda + a^2 = 0$, or $a^2 - b^2$, or $a^2 - c^2$. We notice also that

$$0 < a^2 - b^2 < a^2 - c^2.$$

We proceed now to establish the following theorem.

All the roots of $L_r = 0$ regarded as an equation in $\lambda + a^2$ lie between 0 and $a^2 - c^2$.

Suppose first that $a = b$. Then it has been seen that the function L reduces to the form

$$L = (1 + \tau^2)^{\frac{1}{2}p} D^{p+n} (1 + \tau^2)^n$$

where $\tau^2 = (c^2 + \lambda)/(a^2 - c^2)$, and p has one of the values $0, 1, 2, ..., n$.

The equation $L = 0$ considered as an equation in τ^2 has $\frac{1}{2}p$ or $\frac{1}{2}(p+1)$ coincident roots at $\tau^2 = -1$, according as p is even or odd, together with the roots arising from the polynomial factor $D^{p+n}(1+\tau^2)^n$. To establish the distribution of the roots of this polynomial we can begin with the equation $(1+\tau^2)^n = 0$, and consider the chain of polynomials obtained by successive differentiations. This equation itself, if τ^2 is treated as the variable, has n roots at $\tau^2 = -1$. Again, the derived equation

$$D(1+\tau^2)^n = \tau(1+\tau^2)^{n-1} = 0$$

has $n-1$ roots at $\tau^2 = -1$ and one root at $\tau = 0$. Similarly we get a second derived equation

$$D^2(1+\tau^2)^n = \{1+(2n-1)\tau^2\}(1+\tau^2)^{n-2} = 0.$$

Since $n > 1$ (for the second differentiation to arise), we have that $2n-1 \geqslant 3$, and this equation, in τ^2, has $n-2$ roots at $\tau^2 = -1$ and one root intermediate between -1 and 0.

Continuing the process it is seen that each differentiation produces a further root between -1 and 0, and all these intermediate roots are separate and interlaced with those of the preceding equation of the chain, since they arise one at a time from successive critical values.

The factor $\tau \propto \sqrt{(\lambda+c^2)}$, when present, gives a root at a^2-c^2, but since radical factors are being supposed omitted from L such a root will not occur. Since $1+\tau^2 = (a^2+\lambda)/(a^2-c^2)$, roots $\tau^2 = -1$ correspond to roots $\lambda+a^2 = 0$. Hence, when $a = b$, all the roots of the rational equation

$$L_r(\lambda + a^2) = 0$$

must satisfy $\qquad\qquad 0 \leqslant \text{roots in } \lambda + a^2 < a^2 - c^2,$

and moreover the intermediate roots are necessarily all different.

Before proceeding to show that the same holds good for the roots of the general equation $L_r = 0$, we require to establish the following two simple consequences of Lamé's equation for all a, b, and c.

(i) *Apart from radical factors, $L(\lambda + a^2) = 0$ cannot have a root at 0, a^2-b^2, or a^2-c^2.*

This means that L_r cannot have $\lambda+a^2$, $\lambda+b^2$, or $\lambda+c^2$ as a factor apart from the possible occurrence of these as radicals. To establish this, we have that the general function L satisfies Lamé's equation, which may be written

$$(\lambda+a^2)(\lambda+b^2)(\lambda+c^2)\frac{d^2L}{d\lambda^2} + \tfrac{1}{2}\{(\lambda+b^2)(\lambda+c^2)+(\lambda+c^2)(\lambda+a^2)+(\lambda+a^2)(\lambda+b^2)\}\frac{dL}{d\lambda}$$

$$= (H\lambda+K)L.$$

Suppose now that $L = (\lambda+a^2)f(\lambda)$, where f is a polynomial that does not vanish when $\lambda+a^2 = 0$. Then we have

$$\frac{dL}{d\lambda} = f+(\lambda+a^2)f' = f \qquad \text{for} \quad \lambda+a^2 = 0,$$

$$\frac{d^2L}{d\lambda^2} = 2f'+(\lambda+a^2)f'' = 2f' \qquad \text{for} \quad \lambda+a^2 = 0.$$

Inserting these in the equation and putting $\lambda + a^2 = 0$ gives

$$\tfrac{1}{2}(\lambda + b^2)(\lambda + c^2)f = 0 \qquad \text{for} \quad \lambda + a^2 = 0,$$

and hence f must contain $\lambda + a^2$ as a factor. But this is contrary to hypothesis, and hence L itself cannot contain $\lambda + a^2$ as factor. In exactly the same way it follows that neither $\lambda + b^2$ nor $\lambda + c^2$ can be a factor of L.

Similarly, if we suppose $L = (\lambda + a^2)^{3/2} f(\lambda)$, or suppose $L = (\lambda + a^2)^j f(\lambda)$ where $j \geqslant 1$, and differentiate Lamé's equation an appropriate number of times before substituting for the derivatives of L and then putting $\lambda + a^2 = 0$, we arrive at a similar contradiction.

(ii) *The equation $L = 0$ cannot have a double, or multiple root.*

For, supposing first a double root, then at this value of λ we should have $L = 0$ and $\dfrac{dL}{d\lambda} = 0$, and hence by Lamé's equation

$$(\lambda + a^2)(\lambda + b^2)(\lambda + c^2)\frac{d^2 L}{d\lambda^2} = 0.$$

But, by (i), such a root certainly cannot occur at $-a^2$, $-b^2$, or $-c^2$, and hence at the root $\dfrac{d^2 L}{d\lambda^2} = 0$. This would mean that it must be a triple root, or higher, and is contrary to hypothesis. Hence no root can be double.

If we assume a triple, or higher order root, then by differentiating Lamé's equation an appropriate number of times first we can again arrive at a contradiction.

We can now return to the equation $L_r(\lambda + a^2) = 0$ for the general case $a \neq b$ and prove that all the roots are real and different and lie between 0 and $a^2 - c^2$.

For suppose that the equation has one root, or more, actually greater than $a^2 - c^2$. If then we allow b to change continuously so that it approaches equality with a, this root would at some stage have to pass through the value $\lambda + a^2 = a^2 - c^2$ since it has been shown that for $a = b$ all the roots are less than $a^2 - c^2$. But by (i) no such root can exist, and accordingly a root greater than $a^2 - c^2$ cannot disappear in this way. Similarly there can be no root less than 0.

There is alternatively the possibility, if the equation $L_r = 0$ had two roots greater than $a^2 - c^2$, that these might disappear and become unreal at some stage as b approaches a. But, by continuity, two such roots would first have to become coincident before becoming complex or imaginary, and this again is not possible, by (ii).

Since when $a = b$ all the roots of $L_r = 0$ are real and different it follows that the same must be true when $a \neq b$. Thus we may finally conclude that all the roots of

$$L_r(\lambda + a^2) = 0$$

are real and different, and are intermediate between 0 and $a^2 - c^2$.

ALTERNATIVE METHOD OF CONSTRUCTION
OF ELLIPSOIDAL HARMONICS

The arrangement of the zeros of the several Lamé functions of a given kind can be specified even more closely than has so far been demonstrated, and it is necessary in connexion with the question of the stability of the Jacobi ellipsoids for this more detailed location of the roots to be effected. For this purpose we shall establish an important theorem due to Stieltjes, but the proof of this requires as a preliminary the consideration of the direct construction of ellipsoidal harmonics in rectangular coordinates.

Construction of harmonics of the first species

Harmonics of the first kind can be put in the form

$$Q_1 Q_2 \dots Q_m = \prod_1^m \left(\frac{x^2}{a^2 + \lambda_r} + \frac{y^2}{b^2 + \lambda_r} + \frac{z^2}{c^2 + \lambda_r} - 1 \right).$$

For example, to find the second-order harmonics of this type,

$$\nabla^2 Q = \nabla^2 \left(\frac{x^2}{a^2 + \lambda} + \frac{y^2}{b^2 + \lambda} + \frac{z^2}{c^2 + \lambda} - 1 \right)$$

$$= 2 \left(\frac{1}{a^2 + \lambda} + \frac{1}{b^2 + \lambda} + \frac{1}{c^2 + \lambda} \right),$$

and hence Q will be a spherical harmonic provided λ is a root of the equation

$$(\lambda + b^2)(\lambda + c^2) + (\lambda + c^2)(\lambda + a^2) + (\lambda + a^2)(\lambda + b^2) = 0.$$

This is a quadratic equation with roots λ_1, λ_2, say, such that

$$-a^2 < \lambda_1 < -b^2 < \lambda_2 < -c^2$$

or
$$0 < \lambda_1 + a^2 < a^2 - b^2 < \lambda_2 + a^2 < a^2 - c^2.$$

There are thus two ellipsoidal harmonics of this kind corresponding to these two different possible values of λ.

Next let us consider the conditions for

$$\Pi = Q_1 Q_2 Q_3 \dots Q_m$$

to be harmonic. By direct differentiation we have

$$\frac{\partial \Pi}{\partial x} = \sum_{p=1}^m \frac{\partial \Pi}{\partial Q_p} \cdot \frac{2x}{a^2 + \lambda_p},$$

$$\frac{\partial^2 \Pi}{\partial x^2} = \sum_1^m \frac{\partial \Pi}{\partial Q_p} \cdot \frac{2}{a^2 + \lambda_p} + \sum_{p \neq q} \sum \frac{\partial^2 \Pi}{\partial Q_p \partial Q_q} \cdot \frac{8x^2}{(a^2 + \lambda_p)(a^2 + \lambda_q)}.$$

Hence
$$\nabla^2 \Pi = \sum_1^m \frac{\partial \Pi}{\partial Q_p} \left\{ \frac{2}{a^2 + \lambda_p} + \frac{2}{b^2 + \lambda_p} + \frac{2}{c^2 + \lambda_p} \right\}$$

$$+ \sum_{p \neq q} \sum \frac{\partial^2 \Pi}{\partial Q_p \partial Q_q} \left\{ \frac{8x^2}{(a^2 + \lambda_p)(a^2 + \lambda_q)} + \frac{8y^2}{(b^2 + \lambda_p)(b^2 + \lambda_q)} + \frac{8z^2}{(c^2 + \lambda_p)(c^2 + \lambda_q)} \right\}.$$

But by partial fractions

$$\sum_{\substack{x, y, z \\ a, b, c}} \frac{x^2}{(a^2 + \lambda_p)(a^2 + \lambda_q)} = -\frac{Q_p - Q_q}{\lambda_p - \lambda_q},$$

and also

$$Q_s \frac{\partial^2 \Pi}{\partial Q_s \partial Q_q} = \frac{\partial \Pi}{\partial Q_q},$$

summation over s not being implied. Using these relations to remove the second derivatives of Π we obtain

$$\nabla^2 \Pi = \sum_{p=1}^{m} \frac{\partial \Pi}{\partial Q_p} \left\{ \frac{2}{a^2 + \lambda_p} + \frac{2}{b^2 + \lambda_p} + \frac{2}{c^2 + \lambda_p} + \sum_{q=1}^{m}{}' \frac{8}{\lambda_p - \lambda_q} \right\},$$

where Σ' means, as usual, that the term for which $p = q$ is omitted from the sum. It follows therefore that Π will be harmonic if $\lambda_1, \lambda_2, \ldots, \lambda_m$ are chosen so that each of the separate m quantities in this sum is itself zero. That is, if $\lambda_1, \lambda_2, \ldots, \lambda_m$ are given by a solution of the set of m equations

$$\left. \begin{aligned}
\frac{1}{a^2 + \lambda_1} + \frac{1}{b^2 + \lambda_1} + \frac{1}{c^2 + \lambda_1} + \sum_{q=1}^{m}{}' \frac{4}{\lambda_1 - \lambda_q} &= 0, \\[2mm]
\frac{1}{a^2 + \lambda_2} + \frac{1}{b^2 + \lambda_2} + \frac{1}{c^2 + \lambda_2} + \sum_{q}{}' \frac{4}{\lambda_2 - \lambda_q} &= 0, \\[2mm]
\cdots \quad\quad \cdots \quad\quad \cdots \quad\quad \cdots \quad\quad \cdots \ &= 0, \\[2mm]
\frac{1}{a^2 + \lambda_m} + \frac{1}{b^2 + \lambda_m} + \frac{1}{c^2 + \lambda_m} + \sum_{q}{}' \frac{4}{\lambda_m - \lambda_q} &= 0.
\end{aligned} \right\} \quad (28)$$

These equations, not being linear, will in general have a number of different solutions, and to each solution will correspond an ellipsoidal harmonic of the assumed form.

Let us consider now the polynomial $L(\lambda)$ of degree m in λ defined by

$$L(\lambda) = \prod_{q=1}^{m} (\lambda - \lambda_q),$$

which we shall show to be the Lamé polynomial associated with the foregoing ellipsoidal harmonic. Denoting by dashes differentiation with respect to λ, we

have $L'(\lambda) = \Sigma$ products of $\lambda - \lambda_1, \lambda - \lambda_2, \ldots$ taken $m - 1$ together,

and $L''(\lambda) = 2\Sigma$ products of $\lambda - \lambda_1, \lambda - \lambda_2, \ldots$ taken $m - 2$ together.

Hence, putting $\lambda = \lambda_p$, it is seen that $L''(\lambda_p)/L'(\lambda_p)$ is equal to twice the sum of the reciprocals of $\lambda_p - \lambda_1, \lambda_p - \lambda_2, \ldots, \lambda_p - \lambda_m$, the term $\lambda_p - \lambda_p$ not occurring, of course. That is

$$\frac{L''(\lambda_p)}{L'(\lambda_p)} = 2 \sum_{q=1}^{m}{}' \frac{1}{\lambda_p - \lambda_q} \qquad (q \neq p).$$

Hence if λ has any of the special values $\lambda_1, \lambda_2, \ldots \lambda_p$ given by equations (28) that make Π harmonic, the expression

$$\frac{1}{a^2 + \lambda} + \frac{1}{b^2 + \lambda} + \frac{1}{c^2 + \lambda} + 2 \frac{L''(\lambda)}{L'(\lambda)}$$

vanishes. Multiplying through by the factors involved in the denominators, this means, since L itself is a polynomial, that the expression

$$(a^2+\lambda)(b^2+\lambda)(c^2+\lambda)L''(\lambda)+\tfrac{1}{2}\left\{\sum_{a,\,b,\,c}(b^2+\lambda)(c^2+\lambda)\right\}L'(\lambda)$$

is a polynomial in λ that vanishes whenever λ has any of the special values $\lambda_1, \lambda_2, ..., \lambda_m$. Accordingly it must possess $\lambda-\lambda_1, \lambda-\lambda_2, ..., \lambda-\lambda_m$ as factors, and therefore has the polynomial L itself as a factor. Moreover, it is of degree $m+1$ in λ, and the coefficient of the highest term is clearly

$$m(m-1)+\tfrac{1}{2}.3m = m(m+\tfrac{1}{2}).$$

Hence the remaining factor must be of the form

$$m(m+\tfrac{1}{2})\lambda+K$$

or, if $m=\tfrac{1}{2}n$ $$\tfrac{1}{4}n(n+1)\lambda+K,$$

so that L is simply a polynomial solution of the differential equation

$$(a^2+\lambda)(b^2+\lambda)(c^2+\lambda)L''+\tfrac{1}{2}\{\Sigma(b^2+\lambda)(c^2+\lambda)\}L' = \{\tfrac{1}{4}n(n+1)\lambda+K\}L,$$

which is precisely Lamé's equation.

Harmonics of other species

An exactly similar procedure of direct differentiation can be applied for functions of the three remaining species with the same final result, but with slight changes in the equations corresponding to (27).

The Lamé function of general species can be written

$$(\lambda+a^2)^{\kappa_1}(\lambda+b^2)^{\kappa_2}(\lambda+c^2)^{\kappa_3}\prod_{p=1}^{m}(\lambda-\lambda_p), \tag{29}$$

where the indices $\kappa_1, \kappa_2, \kappa_3$ are always either 0 or $\tfrac{1}{2}$. Also, it has already been shown that $\lambda_1, \lambda_2, ..., \lambda_m$ are all real and unequal, and different from $-a^2, -b^2, -c^2$. The degree, n, of the corresponding harmonic is given by

$$\tfrac{1}{2}n = m+\kappa_1+\kappa_2+\kappa_3,$$

and when $\kappa_1, \kappa_2, \kappa_3$ are specified, the number of Lamé functions of the particular form is settled and always equal to $m+1$. (Table III.)

If we follow out the same procedure of differentiation to find the necessary conditions for

$$\left\{\begin{matrix} & x & yz & \\ 1 & y & zx & xyz \\ & z & xy & \end{matrix}\right\}\prod_{1}^{m}\left(\frac{x^2}{a^2+\lambda_p}+\frac{y^2}{b^2+\lambda_p}+\frac{z^2}{c^2+\lambda_p}-1\right)$$

to be a solution of $\nabla^2 V = 0$, the equations corresponding to (28) in all cases are found to take the form

$$\frac{\kappa_1+\tfrac{1}{4}}{a^2+\lambda_p}+\frac{\kappa_2+\tfrac{1}{4}}{b^2+\lambda_p}+\frac{\kappa_3+\tfrac{1}{4}}{c^2+\lambda_p}+\sum_{q=1}^{m}{}'\frac{1}{\lambda_p-\lambda_q} = 0, \tag{30}$$

wherein $\kappa_1, \kappa_2, \kappa_3$ have the values 0 or $\frac{1}{2}$ as occurring in (29), and the appropriate sets of λ_p's are given by solutions of these equations.

We are now able to establish Stieltjes' theorem on the positions and distributions of the zeros of the Lamé functions of a given order and species.

Stieltjes' theorem

For given $\kappa_1, \kappa_2, \kappa_3$, *the* $m+1$ *Lamé functions* $L(\lambda)$ *of a given order can be arranged in a sequence in such a way that the rational part of the r-th function of the set has* $r-1$ *of its zeros between* $-a^2$ *and* $-b^2$ *and the remaining* $m-r+1$ *zeros between* $-b^2$ *and* $-c^2$.

To prove this, let $\phi_1, \phi_2, ..., \phi_m$ be real variables, m in number, such that

$$-a^2 \leqslant \phi_p \leqslant -b^2 \quad \text{for} \quad p = 1, 2, ..., r-1,$$

and

$$-b^2 \leqslant \phi_p \leqslant -c^2 \quad \text{for} \quad p = r, r+1, ..., m$$

$$(31)$$

and let us consider the following product

$$P = \prod_{p=1}^{m} \left\{ |a^2 + \phi_p|^{\kappa_1 + \frac{1}{2}} |b^2 + \phi_p|^{\kappa_2 + \frac{1}{2}} |c^2 + \phi_p|^{\kappa_3 + \frac{1}{2}} \right\} \prod_{p \neq q} |(\phi_p - \phi_q)|.$$

The function P then has these properties. It is zero when any one or more of the variables ϕ_p have their least values, and also when they have their greatest values. It is essentially positive when the variables ϕ_p are all unequal and also unequal to any of $-a^2$, $-b^2$, or $-c^2$. Furthermore P is a continuous and bounded function in the domain specified by (31), since each of its separate factors is bounded and continuous over the ranges concerned. It follows that there is at least one set of values of the variables, within the domain, at which P attains its upper bound; that is, there exists a set of values within the domain for which P, and therefore $\log P$, has a maximum value. These values of $\phi_1, \phi_2, ..., \phi_m$ will be given by the equations

$$\frac{\partial \log P}{\partial \phi_1} = \frac{\partial \log P}{\partial \phi_2} = ... = \frac{\partial \log P}{\partial \phi_m} = 0.$$

But these equations, when written out, are precisely equations (30), with ϕ_p occurring instead of λ_p. Hence each solution of (30) necessarily lies *within* the domain (31), and not on its boundary.

Thus if r has any of the values $1, 2, ..., m+1$, there exists a corresponding Lamé function whose rational part is such that $r-1$ of its zeros lie between $-a^2$ and $-b^2$ and the remaining $m-r+1$ lie between $-b^2$ and $-c^2$. Moreover, since when $\kappa_1, \kappa_2, \kappa_3$ are specified, there are precisely $m+1$ Lamé functions of the given species, it follows that they are all obtained in turn when r is given successively the values $1, 2, ..., m+1$.

We notice finally that if $r = 1$ there are no roots of $L(\lambda)$ between $-a^2$ and $-b^2$, while if $r = m+1$ there are no roots between $-b^2$ and $-c^2$.

Chapter VI

FURTHER PROPERTIES OF LAMÉ FUNCTIONS AND APPLICATION TO GRAVITATION

The ellipsoid, E,
$$\frac{x^2}{a^2}+\frac{y^2}{b^2}+\frac{z^2}{c^2}=1 \tag{1}$$

can be regarded as the surface $\lambda = 0$, and any point of its surface (in the positive quadrant) can be specified by coordinates (μ, ν). The perpendicular distance from the origin to the tangent plane at the point (μ, ν) is easily found to be

$$\frac{abc}{\sqrt{(\mu\nu)}}.$$

It will be convenient to denote this perpendicular distance by ϖabc so that we have
$$\varpi = 1/\sqrt{(\mu\nu)},$$

and is necessarily always finite and positive. This quantity is of special importance in establishing the following so-called orthogonal property of the surface harmonics associated with an ellipsoid.

If MN and $M_1 N_1$ are two different surface harmonics, then

$$\iint_E \varpi MN M_1 N_1 \, dS = 0, \tag{2}$$

where dS is an element of the surface (1), and the double integral is extended over the whole surface of the ellipsoid (E).

This result can readily be established by means of Green's theorem. For consider the two solid harmonics related to these surface harmonics, namely

$$V = LMN \quad \text{and} \quad V_1 = L_1 M_1 N_1.$$

Then $\nabla^2 V = 0$ and $\nabla^2 V_1 = 0$ everywhere. But by Green's theorem

$$\iiint (V_1 \nabla^2 V - V \nabla^2 V_1) \, dx \, dy \, dz = \iint_E \left(V \frac{\partial V_1}{\partial n} - V_1 \frac{\partial V}{\partial n} \right) dS,$$

where the volume integral is taken through the interior of the ellipsoid and the surface integral over its surface, and $\dfrac{\partial}{\partial n}$ denotes differentiation along the outward normal. Since V and V_1 are harmonic, the left-hand side is identically zero and the surface integral on the right must accordingly vanish.

Now for an infinitesimal step along the normal only the coordinate λ changes, and hence

$$\frac{\partial V}{\partial n} = MN \frac{\partial L}{\partial n} = MN \frac{dL}{d\lambda} \cdot \frac{\partial \lambda}{\partial n}.$$

But by equations (13), Chapter V, p. 54,

$$dn = h_1 d\lambda = \frac{\sqrt{[(\lambda-\mu)(\lambda-\nu)]}}{2\sqrt{[(a^2+\lambda)(b^2+\lambda)(c^2+\lambda)]}} d\lambda$$

$$= \frac{\sqrt{(\mu\nu)}}{2abc} d\lambda \quad \text{on the surface } \lambda = 0$$

$$= \frac{1}{2\varpi abc} d\lambda.$$

The surface integral thus becomes

$$\iint_E MNM_1N_1\left(L\frac{dL_1}{d\lambda} - L_1\frac{dL}{d\lambda}\right)\varpi abc\, dS = 0.$$

But the factor $L\frac{dL_1}{d\lambda} - L_1\frac{dL}{d\lambda}$ depends only on λ, and therefore remains constant over (E), or over any λ-surface, and may thus be taken outside the integral. Also, this factor cannot be zero, for if it were we would have

$$\frac{1}{L_1}\frac{dL_1}{d\lambda} = \frac{1}{L}\frac{dL}{d\lambda},$$

or

$$L_1 = kL,$$

where k is a constant, which would mean that L and L_1 were not essentially different Lamé functions. The orthogonality property (2) is accordingly proved.

It is important to remember that the result holds not only when MN and M_1N_1 are of different orders (n), but also for any two different harmonics of the same order.

Development of a function as a series of ellipsoidal surface harmonics

It has been seen that any continuous function of position (μ,ν) on the surface of an ellipsoid can be expressed as a linear sum of the surface harmonics MN, and the foregoing orthogonality property now enables the coefficients in the expansion to be formally obtained.

Thus for any function $\Phi(\mu,\nu)$ we must have an expansion of the form

$$\Phi = \sum_k A_k M_k N_k,$$

where the A_k are certain constants. Multiplying through by $\varpi M_k N_k$ and integrating over the surface of the ellipsoid gives

$$\iint_E \varpi\Phi(\mu,\nu)M_kN_k\,dS = \iint_E \varpi\left(\sum_i A_i M_i N_i\right)M_kN_k\,dS,$$

and all the products of different harmonics on the right vanish when integrated except that for which $i = k$. Whence

$$\iint_E \varpi\Phi M_kN_k\,dS = A_k\iint_E \varpi M_k^2 N_k^2\,dS, \tag{3}$$

and this relation formally determines the constants A_k of the ellipsoidal (surface) harmonic expansion of Φ. It will be convenient to term them the *expansion constants* of the function $\Phi(\mu, \nu)$. It is to be noticed that the factor multiplying A_k on the right in (3) is an essentially positive quantity.

The linear independence of the $2n+1$ ellipsoidal harmonics of a given order n

This can now be easily established as a simple consequence of the orthogonality. For suppose $L_s M_s N_s$ ($s = 1, 2, \ldots, 2n+1$) are the ellipsoidal harmonics of a given order n, and let it be assumed that $2n+1$ constants c_s exist, not all zero, such that

$$\sum_1^{2n+1} c_s L_s M_s N_s \equiv 0.$$

If we multiply this by $\varpi M_k N_k$ where $M_k N_k$ is any selected one of the $2n+1$ ellipsoidal surface harmonics of order n, and integrate over the surface of the ellipsoid, we get

$$c_k L_k \iint \varpi M_k^2 N_k^2 dS = 0,$$

all other terms disappearing because of the orthogonal property. But the integral is essentially positive, so that we must have $c_k = 0$ for all k, and this is contrary to the original supposition. Accordingly no such linear relation can exist, and the ellipsoidal harmonics of any given order are therefore linearly independent.

The function $S(\lambda)$ associated with L

Since Lamé's equation is a second order differential equation there must exist two independent solutions for each value of K. For given appropriate K the function L can be regarded as one of these, and there must then exist another independent solution which will be denoted by S.

To determine this second solution, suppose the independent variable λ is replaced in Lamé's equation by u where

$$d\lambda = - 2 \sqrt{[(a^2 + \lambda)(b^2 + \lambda)(c^2 + \lambda)]} du = - 2A du.$$

Then
$$A \frac{d}{d\lambda} \left(A \frac{d}{d\lambda} \right) = \frac{1}{4} \frac{d^2}{du^2},$$

and Lamé's equation may be written

$$\frac{d^2 L}{du^2} = 4(H\lambda + K)L.$$

The function S will satisfy the same equation, so that

$$\frac{d^2 S}{du^2} = 4(H\lambda + K)S.$$

Multiplying the first of these equations by S and the second by L, and subtracting, gives

$$L \frac{d^2 S}{du^2} - S \frac{d^2 L}{du^2} = 0,$$

or, after integration,
$$L \frac{dS}{du} - S \frac{dL}{du} = \text{constant}.$$

If this constant of integration were taken as zero, a further integration would give $S = kL$, as seen above. This must, of course, be a possible solution, but it is obviously not a new function. If, however, the constant is taken to be other than zero, S then turns out to be a new function altogether. The actual value taken for the constant, so long as it is not zero, is immaterial, since changing it can only multiply S by a constant factor. It turns out to be convenient to choose it to be $2n + 1$ where n is the order of the harmonic LMN. We thus have

$$\frac{1}{L^2}\left(L\frac{dS}{du} - S\frac{dL}{du}\right) = \frac{2n+1}{L^2}$$

which integrates to give

$$\frac{S}{L} = \int_0^u \frac{2n+1}{L^2}\,du,$$

or, in terms of λ,

$$S(\lambda) = L\int_\lambda^\infty \frac{(2n+1)\,d\lambda}{2L^2\sqrt{[(a^2+\lambda)(b^2+\lambda)(c^2+\lambda)]}} \tag{4}$$

the upper limit of integration again being chosen with a view to later convenience.

The function $S(\lambda)$ found in this way is therefore such that

$$S(\lambda)\,M(\mu)\,N(\nu)$$

also satisfies Laplace's equation in curvilinear coordinates λ, μ, ν.

It will be seen that $S \to 0$ as $\lambda \to \infty$. For λ large, we have $L = \alpha\lambda^{\frac{1}{2}n} = \alpha r^n$, where α is any constant factor, and also $A \to \lambda^{3/2}$. Hence, for λ large

$$S = \alpha\lambda^{\frac{1}{2}n}\int_\lambda^\infty \frac{2n+1}{a^2\lambda^n}\cdot\frac{d\lambda}{2\lambda^{3/2}}$$

$$= \alpha^{-1}\lambda^{-\frac{1}{2}(n+1)} = \alpha^{-1}r^{-(n+1)}.$$

Thus, in the analogy with spherical harmonics, the function SMN corresponds to functions of the type $r^{-n-1}P_n^p(\cos\theta)\begin{smallmatrix}\cos\\\sin\end{smallmatrix}p\phi$.

It thus appears that the general normal solution of Laplace's equation in terms of λ, μ, ν is of the form

$$\Sigma\left\{\alpha_1 L(\lambda) + \beta_1 L\int_\lambda \frac{d\lambda}{L^2 A}\right\}\left\{\alpha_2 M + \beta_2 M\int_\mu \frac{du}{M^2 B}\right\}\left\{\alpha_3 N + \beta_3 N\int_\nu \frac{dv}{N^2 C}\right\}$$

corresponding to the function

$$\Sigma(\alpha_1 r^n + \beta_1 r^{-n-1})\{\alpha_2 P_n^p(\cos\theta) + \beta_2 Q_n^p(\cos\theta)\}(\alpha_3\cos p\phi + \beta_3\sin p\phi)$$

of ordinary spherical harmonics.

It may be noted also that as $\lambda \to \infty$

$$LS \to \alpha\lambda^{\frac{1}{2}n}.\alpha^{-1}\lambda^{-\frac{1}{2}(n+1)} = \lambda^{-\frac{1}{2}}$$

independently of n and α.

The Dirichlet problem for an ellipsoid

Suppose we are given the values that a function takes on the surface of the ellipsoid (1) and we wish to find a harmonic function V defined in all space that takes these values on the ellipsoid.

Let $\Phi(\mu, \nu)$ denote the given function, to which V must reduce on the ellipsoid. It must be expressible as a linear sum of surface harmonics in the form

$$\Phi = \Sigma A_1 M_1 N_1.$$

We can replace each constant A_1 in this by a slightly different constant by putting

$$A_i = \alpha_i L_i(\lambda=0)\, S_i(\lambda=0),$$

where L_i is the Lamé function in λ associated with $M_i N_i$, and S_i is the corresponding function. We thus have, on the ellipsoid

$$\Phi = \Sigma \alpha_i L_i(0)\, S_i(0)\, M_i N_i,$$

where $L_i(0)$ and $S_i(0)$ mean the values of these functions on $\lambda = 0$.

Now consider the solid harmonic function defined as follows:

$$
\left.
\begin{aligned}
V &= \Sigma \alpha S(0) LMN \quad \text{inside the ellipsoid,} \\
V &= \Sigma \alpha L(0) SMN \quad \text{outside.}
\end{aligned}
\right\}
\tag{5}
$$

This function evidently satisfies $\nabla^2 V = 0$ everywhere, and takes the required values Φ on the surface of the ellipsoid, and accordingly represents the solution of the problem.

The close analogy with the comparable result in spherical harmonic analysis is obvious.

The gravitational potential of a surface layer on the ellipsoid

The above function V is evidently continuous across the surface of the ellipsoid, but its normal gradient is not. To see this, we calculate $\dfrac{\partial V}{\partial n}$ on the two sides of the surface. At points just outside we have, by outward differentiation,

$$
\left(\frac{\partial V}{\partial n}\right)_0 = \Sigma \alpha L(0)\, MN \frac{dS}{du} \cdot \frac{du}{d\lambda} \cdot \frac{\partial \lambda}{\partial n}
$$

$$
= -\Sigma \alpha L(0)\, S' MN \cdot \frac{1}{2A} \cdot 2\varpi\, abc
$$

$$
= -\varpi \Sigma \alpha L(0)\, S'(0)\, MN \quad \text{on the surface } \lambda = 0,
$$

where dashes denote differentiation with respect to u. Similarly at points just inside we have, by inward differentiation,

$$
\left(\frac{\partial V}{\partial n}\right)_i = -\varpi \Sigma \alpha S(0)\, L'(0)\, MN.
$$

If we regard V as the gravitational potential of a surface layer of material of given surface density $\sigma(\mu, \nu)$, then V must be continuous across the surface, but its gradient need not in general be so. Indeed, by Gauss's theorem, considering in the usual way the flux out of a flat elementary cylinder containing the surface element of mass σdS and with its end faces parallel to the surface,

$$\left(\frac{\partial V}{\partial n}\right)_0 - \left(\frac{\partial V}{\partial n}\right)_i = -4\pi\sigma,$$

assuming the constant of gravitation to be unity, or absorbed in σ. Instead of a surface distribution of zero thickness and (variable) surface density σ, it can equally be supposed that V is the potential of a distribution of unit density everywhere but of variable infinitesimal normal depth ξ, and we shall then have a corresponding equation

$$\left(\frac{\partial V}{\partial n}\right)_0 - \left(\frac{\partial V}{\partial n}\right)_i = -4\pi\xi.$$

Accordingly, if we are given a potential function V of the above form, the thickness of the surface distribution giving rise to it will be determined by

$$4\pi\xi = \varpi\Sigma\alpha\{L(0)S'(0) - L'(0)S(0)\}MN,$$

and since L and S are so defined that

$$LS' - L'S = 2n + 1,$$

where n is the order of the L function, we obtain finally

$$4\pi\xi = \varpi\Sigma\alpha(2n + 1)MN.$$

Reversing the process, if instead we are given a surface distribution of unit density and depth ξ of the form

$$\xi = \varpi\Sigma\beta_i M_i N_i,$$

then the potential at all points of space can be written down at once as

$$\left.\begin{array}{l} V(\text{outside}) = \Sigma\dfrac{4\pi\beta_i}{2n+1}L_i(0)S_i M_i N_i, \\[3mm] V(\text{inside}) \;\;= \Sigma\dfrac{4\pi\beta_i}{2n+1}S_i(0)L_i M_i N_i, \end{array}\right\} \tag{6}$$

and the value of the potential on the surface is

$$\Phi = \Sigma\frac{4\pi\beta_i L_i(0)S_i(0)}{2n+1}M_i N_i.$$

Liouville's formula

We can now establish a certain important relation of which we shall have need later.

If we are given a surface distribution

$$\Sigma\varpi\beta_i M_i N_i,$$

its potential at a general point P of space can be written down directly. In fact, if Q denotes the distance of the element dS from P, then

$$V(P) = \iint_E \Sigma \varpi \beta_i M_i N_i \frac{dS}{Q}$$

and this must be equal to one or other of expressions (6) according as P is outside or inside the surface. Since, however, the β_i may be regarded as arbitrary, we can suppose them all zero except a particular one, and the equality of the two forms for the potential becomes

$$\iint_E \varpi M N Q^{-1} dS = \frac{4\pi}{2n+1} L(0) SMN$$

if P is outside the ellipsoid, and

$$\iint_E \varpi M N Q^{-1} dS = \frac{4\pi}{2n+1} S(0) LMN$$

if P is inside.

If now P is allowed to become a point P_0 of the surface $\lambda = 0$, the relation reduces to Liouville's formula, namely

$$\iint_E \varpi M N Q^{-1} dS = \frac{4\pi}{2n+1} L(0) S(0) MN$$

where Q now denotes the distance of dS from the point P_0 of the surface itself.

Use of constant angular velocity for a freely rotating mass in stability considerations

We return now to the question raised in Chapter II (p. 27), and show that for certain disturbances or deformations of a freely rotating ellipsoidal liquid mass, it is permissible to examine the stability as if ω remained constant. That is, the equations can be set up by using axes rotating with constant ω equal to the value in the equilibrium configuration.

For, consider a surface distribution of (small) depth ξ measured normal to the surface of the ellipsoid (ξ is of course a negative quantity at some parts of the surface). In the case that will concern us most, ξ/ϖ is a third-order surface harmonic. The change in moment of inertia of the system, δI, can be found simply by calculating the moment of inertia of the surface layer, and this is

$$\delta I = \int_E \int_0^\xi (\varrho + h \cos \phi)^2 \, dh \, dS$$

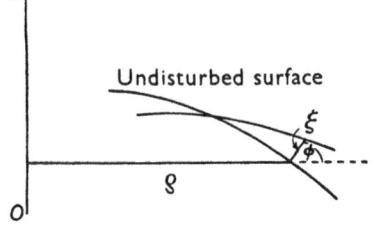

Fig. 14.

where ϱ is the distance of the surface element dS of the ellipsoid (E) from the axis of rotation Oz, h is the elevation above the surface of an element of volume $dh\,dS$, and ϕ is the angle between the radial direction, which is that of ϱ, and the normal at dS.

Thus
$$\delta I = \int_E \int_0^\xi (\varrho^2 + 2\varrho h \cos\phi + h^2 \cos^2\phi)\, dh\, dS$$
$$= \int_E (\varrho^2 \xi + \varrho \xi^2 \cos\phi + \tfrac{1}{3}\xi^3 \cos^2\phi)\, dS.$$

Now $\varrho^2 = x^2 + y^2$, and on the surface of the ellipsoid this is expressible as a linear sum of second-order surface harmonics. For, in the solid expression

$$x^2 + y^2 + \theta\left(\frac{x^2}{a^2} + \frac{y^2}{b^2} + \frac{z^2}{c^2} - 1\right),$$

θ can be chosen so that the function is harmonic, namely, so that

$$2 + \theta\left(\frac{1}{a^2} + \frac{1}{b^2} + \frac{1}{c^2}\right) = 0.$$

The quantity ϱ^2 is then expressible in terms of second-order solid ellipsoidal harmonics, that is, as a sum $\Sigma A_i L_i M_i N_i$ for each term of which $n = 2$. On the surface it therefore reduces to $\Sigma \alpha_i M_i N_i$ since λ is constant. Accordingly, $\int_E \varrho^2 \xi\, dS = 0$, since the integrand is the product of second-order and third-order harmonics together with the factor ϖ. Thus δI contains terms involving only ξ^2 and higher powers, and is therefore of second order of smallness, which may be indicated by writing
$$I = I_0 + \delta_2 I.$$

But since the total angular momentum is invariable $\delta(I\omega) = 0$ and hence

$$\delta\omega = -\frac{\omega}{I}\,\delta I,$$

and the change in angular velocity is therefore also of second order.

Now in considering *secular stability* we are concerned with the second-order variation of $W + M^2/2I$; the first-order change is already zero since the deformations are from an equilibrium form. We thus have

$$\delta\left(W + \frac{M^2}{2I}\right) = \delta_2 W - \frac{M^2}{2I^2}\delta_2 I \quad \text{to second order}$$
$$= \delta_2 W - \tfrac{1}{2}\omega^2 \delta_2 I,$$

correct to the second order of smallness. Hence the conditions will be exactly the same as for the function $W - \tfrac{1}{2}\omega^2 I$ with ω treated as constant.

The same will evidently also be true in discussing *ordinary stability* if the deformation involved in the oscillations is expressible by third-order harmonics or higher. For the equations of small motion consist of terms of first order of smallness, and hence any additional terms that would arise through $\delta\omega$, which has been seen is necessarily second order, can be neglected as of the second order of smallness. Thus constant angular velocity may again be assumed.

We may notice that these results will also apply to certain second-order harmonic deformations provided they are such as not to affect the moment of inertia to the first order of small quantities.

Calculation of surface gravity on the ellipsoid

We next establish the important result that on the surface of an equilibrium ellipsoidal figure the value of apparent gravity, g, at any point is inversely proportional to the perpendicular distance from the centre to the tangent plane at the point. That is, $g\varpi$ is constant over the surface.

To prove this, we have that at points within the ellipsoid the total mechanical potential, that is, gravitational plus centrifugal, may be written

$$U = -\pi(\alpha x^2 + \beta y^2 + \gamma z^2) + \tfrac{1}{2}\omega^2(x^2 + y^2) + \text{constant},$$

where α, β, γ are defined by equations (4) of Chapter IV, and we have taken $G\rho = 1$ for convenience, as is permissible. The components of apparent gravity within and on the surface, measured parallel to the coordinate axes, are

$$\frac{\partial U}{\partial x}, \; \frac{\partial U}{\partial y}, \; \frac{\partial U}{\partial z}.$$

But in equilibrium g is along the inward normal to the ellipsoid, whose direction cosines are

$$-\frac{\varpi x}{a^2}, \; -\frac{\varpi y}{b^2}, \; -\frac{\varpi z}{c^2}.$$

Hence considering the third component we have

$$g \cdot \frac{\varpi z}{c^2} = -\frac{\partial U}{\partial z} = 2\pi\gamma z.$$

Accordingly the product $g\varpi$ is constant and is given by

$$g\varpi = 2\pi c^2 . abc \int_0^\infty \frac{d\lambda}{(c^2 + \lambda)\sqrt{[(a^2 + \lambda)(b^2 + \lambda)(c^2 + \lambda)]}}. \tag{7}$$

The result can be expressed more briefly, and also in a form specially suited to our purpose, in terms of Lamé functions. For, the first-order ($n = 1$) Lamé function associated with the *third* axis c of the ellipsoid, namely,

$$\sqrt{(c^2 + \lambda)} = L_1, \quad \text{say},$$

has corresponding S-function given by

$$S_1 = \sqrt{(c^2 + \lambda)} \int_\lambda^\infty \frac{3d\lambda}{2(c^2 + \lambda)\sqrt{[(a^2 + \lambda)(b^2 + \lambda)(c^2 + \lambda)]}}.$$

These two functions L_1 and S_1 are of course functions of λ, but on the surface $\lambda = 0$ we have at once that

$$g\varpi = \tfrac{4}{3}\pi abc L_1(0) S_1(0),$$

but it will be convenient to write this result for brevity simply as

$$g\varpi = \tfrac{4}{3}\pi L_1 S_1,$$

wherein it is to be understood that the Lamé functions are to be evaluated on the surface of the ellipsoid $\lambda = 0$, and the product of the axes has been taken as unity, that is, $abc = 1$.

Calculation of the coefficients of stability of an ellipsoidal configuration

With the results of this chapter available we can now obtain the change of potential energy (gravitational and centrifugal) of a rotating ellipsoidal mass of liquid when its free surface is subjected to a general small deformation.

The undisturbed form of the surface has equation $\lambda = 0$ in ellipsoidal co-ordinates, and any point of it has coordinates (μ, ν). The motion is a pure rotation round Oz. We now suppose the surface to undergo an infinitesimal continuous displacement without change of total volume of figure, and we denote by ξ the distance from a point of the original ellipsoidal surface to the deformed surface measured along the normal to the ellipsoid. As before, we denote by g the resultant force at (μ, ν) due to the gravitational attraction and the centrifugal force. Also, we denote by dS an element of surface area surrounding the point (μ, ν).

The total force acting on a particle of matter in the deformed configuration may be regarded as made up of three contributions, thus:

(i) the attraction of the original equilibrium ellipsoid,

(ii) the attraction of the surface layer formed by the difference, plus and minus, between the deformed surface and the ellipsoid,

(iii) the centrifugal force.

If dm_E denotes an element of mass of the ellipsoid, and Q its distance from a given arbitrary point P, the total mechanical potential at P due to (i) and (iii) is

$$V = -\int \frac{dm_E}{Q} - \tfrac{1}{2}\omega^2 r^2, \tag{8}$$

where r is the distance of P from the axis of rotation. The product of the constant of gravitation and the density, $G\rho$, may be taken as unity.

Since the ellipsoid is, by hypothesis, an equilibrium form, V will be constant over its surface and equal to V_0, say. Also, the gradient of V will be the total force that would act on a unit mass at P at rest in the rotating frame due to gravitation plus rotation. If P is at a *small* distance h along the normal to the equilibrium surface at a point P_1, then to the first order of small quantities

$$V(P) = V_0(P_1) + gh, \tag{9}$$

and here g may be calculated anywhere within first-order distance of P or P_1 without introducing first-order error in $V(P)$.

If now dm and dm' are any two elements of mass at distance Δ apart, after the deformation has been applied, the potential energy due to gravitation will be

$$-\tfrac{1}{2}\iint \Delta^{-1} dm\, dm'$$

and the potential energy due to centrifugal force will be

$$-\tfrac{1}{2}\int \omega^2 r^2\, dm,$$

the integrations being extended to every element of mass of the deformed body, and the sign conventions taken in both cases so that the force is minus the gradient of the potential. Thus, the total potential U, say, is given by

$$2U = -\iint \Delta^{-1} dm\, dm' - \omega^2 \int r^2 dm.$$

If we denote by U_0 the value of U in the undisturbed configuration, then stability depends on the function $U - U_0$. To evaluate this in suitable form, let us denote by dm_E, dm'_E elements of mass of the original ellipsoid, and by dm_L, dm'_L elements of mass of the surface layer. Since the whole mass, and volume, are unchanged, we have always

$$\int dm_L = 0 = \int dm'_L,$$

and

$$\iint \Delta^{-1} dm\, dm' = \iint \Delta^{-1} (dm_E + dm_L)(dm'_E + dm'_L).$$

Hence with this subdivision of the total mass we have

$$-2U = \iint \Delta^{-1} dm_E\, dm'_E + \iint \Delta^{-1} dm_E\, dm'_L + \iint \Delta^{-1} dm'_E\, dm_L + \iint \Delta^{-1} dm_L\, dm'_L$$

$$+ \omega^2 \int r^2 dm_E + \omega^2 \int r^2 dm_L$$

and

$$-2U_0 = \iint \Delta^{-1} dm_E\, dm'_E + \omega^2 \int r^2 dm_E,$$

where in each integral Δ denotes the distance between the elements of mass concerned. But obviously

$$\iint \Delta^{-1} dm_E\, dm'_L = \iint \Delta^{-1} dm'_E\, dm_L = \iint \Delta^{-1} dm_E\, dm_L,$$

and hence

$$-2(U - U_0) = 2 \iint \Delta^{-1} dm_E\, dm_L + \iint \Delta^{-1} dm_L\, dm'_L + \omega^2 \int r^2 dm_L$$

$$= -2 \int V dm_L + \iint \Delta^{-1} dm_L\, dm'_L$$

in virtue of equation (8).

To reduce this further, consider an arbitrary element of area dS of the surface of the ellipsoid. The normals at its boundary generate a cylinder with dS as base, and the volume element dm_L may be defined by taking a thin slice of this cylinder bounded by planes parallel to dS at distances h and $h + dh$ above the surface. We may then write

$$dm_L = dS\, dh, \quad dm'_L = dS'\, dh',$$

and h and h' lie between 0 and ξ (which takes positive and negative values), where ξ is itself a small quantity. We thus have

$$\int V dm_L = \int V_0\, dm_L + \int gh\, dm_L$$

$$= V_0 \int dm_L + \int g\, dS \int_0^\xi h\, dh$$

$$= \tfrac{1}{2} \int g \xi^2\, dS,$$

correct to the second order of small quantities.

Again
$$\iint Q^{-1} dm_L \, dm_L' = \iiiint Q^{-1} dh \, dS \, dh' \, dS',$$

and since ξ and ξ' are first-order small quantities, Q may be taken as the distance between the surface elements dS, dS' of the ellipsoid without affecting the accuracy as far as second order. Thus

$$\iint Q^{-1} dm_L \, dm_L' = \iint Q^{-1} dS \, dS' \int_0^\xi dh \int_0^{\xi'} dh' = \iint Q^{-1} \xi \xi' \, dS \, dS',$$

and we have, correct to the second order,

$$2(U - U_0) = \int g \xi^2 dS - \iint Q^{-1} \xi \xi' \, dS \, dS'. \tag{10}$$

Let us now assume that ξ/ϖ, which is a function of position on the surface of the ellipsoid, consists of a convergent infinite series of the form

$$\xi/\varpi = \sum_i A_i M_i(\mu) N_i(\nu),$$

wherein the A_i's are (infinitesimal) constant coefficients. These A_i may be regarded as the coordinates of the dynamical system, but now infinite in number. Thus if (μ, ν) is a point within dS, and (μ', ν') a point within dS', we have

$$\xi = \sum_i A_i \varpi M_i N_i,$$

$$\xi' = \sum_i A_i \varpi' M_i' N_i',$$

where the dashes indicate that the functions are calculated at (μ', ν').

Thus
$$\xi^2 = \sum_i A_i^2 \varpi^2 M_i^2 N_i^2 + 2 \sum_i \sum_{i \neq k} A_i A_k \varpi^2 M_i N_i M_k N_k,$$

and hence

$$\int g \xi^2 dS = \Sigma A_i^2 \int g \varpi^2 M_i^2 N_i^2 dS + 2 \sum_i \sum_{i \neq k} A_i A_k \int g \varpi^2 M_i N_i M_k N_k dS$$

$$= \tfrac{4}{3} \pi L_1 S_1 \Sigma A_i^2 \int \varpi M_i^2 N_i^2 dS + \tfrac{8}{3} \pi L_1 S_1 \sum_i \sum_{i \neq k} A_i A_k \int \varpi M_i N_i M_k N_k dS,$$

making use of the fact that $g\varpi = \tfrac{4}{3}\pi L_1 S_1$ and is constant over the surface, so that it may be taken outside the sign of integration. But every term in the second double sum is zero on account of the orthogonality property of the surface harmonics, so that finally

$$\int g \xi^2 dS = \tfrac{4}{3} \pi L_1 S_1 \sum_i A_i^2 \int \varpi M_i^2 N_i^2 dS.$$

Inserting the values of ξ and ξ' in the second integral on the right of (10) gives

$$\sum_{i, k} A_i A_k \iint Q^{-1} \varpi \varpi' M_i N_i M_k' N_k' \, dS \, dS',$$

and here the summations are over all i and k, equal or not. But by Liouville's formula

$$\int Q^{-1}\varpi' M'_k N'_k dS' = \frac{4\pi}{2n+1} L_k(0) S_k(0) M_k(\mu) N_k(\nu),$$

where n is the order of the Lamé function L_k. The double integral accordingly becomes

$$\sum_{i,\,k} A_i A_k \frac{4\pi}{2n+1} L_k(0) S_k(0) \int \varpi M_i N_i M_k N_k dS.$$

In this, in any term for which $i \neq k$ the integral vanishes, and the expression reduces to the single summation

$$\sum_i \frac{4\pi}{2n+1} L_i(0) S_i(0) A_i^2 \int \varpi M_i^2 N_i^2 dS.$$

Accordingly we arrive at the final expression for the change of potential energy, namely,

$$U - U_0 = 2\pi \Sigma \left(\frac{1}{3} L_1 S_1 - \frac{1}{2n+1} L_i S_i \right)_{\lambda=0} A_i^2 \int_E \varpi M_i^2 N_i^2 dS. \tag{11}$$

The form of the displaced free surface may be regarded as defined by the (small) coefficients A_i, which correspond to the q_i of ordinary dynamical systems involving a finite number of degrees of freedom. Here, however, the A_i's are infinite in number, there being one corresponding to each Lamé function, so that there are $2n+1$ such coordinates associated, one to one, with the $2n+1$ surface harmonics $M_i N_i$ of a given order n. We have accordingly obtained the expression of U as far as the second order in the coordinates. First-order terms do not occur, of course, since the displacement is from an equilibrium configuration. To each coordinate A_i there corresponds a coefficient of stability

$$4\pi \left(\frac{1}{3} L_1 S_1 - \frac{1}{2n+1} L_i S_i \right)_{\lambda=0} \int_E \varpi M_i^2 N_i^2 dS, \tag{12}$$

and in this coefficient the factor $4\pi \int \varpi M_i^2 N_i^2 dS$ is essentially positive whatever the ratios of the axes $a : b : c$. Hence in order that a given spheroidal or ellipsoidal configuration shall be a form of bifurcation it is necessary that one or more of these coefficients shall vanish for the ratios of axes involved. Thus secular stability is seen to depend on the equation

$$F_i \equiv \frac{1}{3} L_1 S_1 - \frac{1}{2n+1} L_i S_i = 0, \tag{13}$$

which, since $\lambda = 0$, constitutes a (transcendental) relation between the ratios of the axes $a : b : c$. The discussion of its possible roots, first when $a = b$, and second for the Jacobi ellipsoids, forms the subject of the two following chapters. In the subsequent work it will be convenient to ignore the essentially positive factors and call simply F_i the coefficient of stability.

Surface displacements given by first-order harmonics

It will be recalled that the Lamé function L_1 occurring in the coefficients of stability is $\sqrt{(\lambda + c^2)}$ (before λ is put equal to zero). Accordingly the third of the three coefficients of stability corresponding to $n = 1$ is always zero. The reason for this is because the equilibrium is neutral for any small translation of the ellipsoid as a whole parallel to the axis of rotation. To see this, suppose the ellipsoid displaced by δc parallel to Oz, then the equations of the free surface before and after when expressed in the same coordinate system are

$$\frac{x^2}{a^2} + \frac{y^2}{b^2} + \frac{z^2}{c^2} = 1,$$

and

$$\frac{x^2}{a^2} + \frac{y^2}{b^2} + \frac{(z - \delta c)^2}{c^2} = 1.$$

If, as usual, we denote by ξ the normal displacement of a point P' of the second surface from a point $P(x, y, z)$ of the original surface, the coordinates of P' are

$$\left(x + \frac{\varpi x}{a^2}\xi, \quad y + \frac{\varpi y}{b^2}\xi, \quad z + \frac{\varpi z}{c^2}\xi\right),$$

and this lies on the second surface if

$$\frac{1}{a^2}\left(x + \frac{\varpi x}{a^2}\xi\right)^2 + \frac{1}{b^2}\left(y + \frac{\varpi y}{b^2}\xi\right)^2 + \frac{1}{c^2}\left(z + \frac{\varpi z}{c^2}\xi - \delta c\right)^2 = 1.$$

Retaining only the first-order terms gives

$$\xi\varpi\left(\frac{x^2}{a^4} + \frac{y^2}{b^4} + \frac{z^2}{c^4}\right) = \frac{z\delta c}{c^2}.$$

But

$$\frac{x^2}{a^4} + \frac{y^2}{b^4} + \frac{z^2}{c^4} = \frac{1}{\varpi^2} \quad \text{(assuming } abc = 1\text{),}$$

and hence

$$\xi = \frac{\delta c}{c^2}\varpi z.$$

But z is one of the three ellipsoidal harmonics of order 1, and on the surface of the ellipsoid

$$z \propto \sqrt{[(\mu + c^2)(\nu + c^2)]} = M_1(\mu)N_1(\nu);$$

whence we have

$$\xi = A\varpi M_1 N_1,$$

where the coordinate A is a small quantity proportional to δc.

In the same way a small displacement parallel to Ox or Oy also leads to a value of ξ/ϖ proportional to the corresponding first-order harmonic. Such deformations, however, involve a displacement of the centre of gravity of the whole mass off the axis of rotation and are accordingly ruled out since this will not affect the question of the relative stability of figure of the mass as a whole considered quite apart from any possible translatory motion. For this purpose the centre of gravity can always be taken at rest on the axis of rotation.

Chapter VII

THE SECULAR STABILITY OF THE MACLAURIN SPHEROIDS

The determination of the secular stability for all small displacements involves finding the points of bifurcation on the Maclaurin series. We may regard this series as starting at the spherical form, which has already been shown to be thoroughly stable, and developing in the direction of increasing angular momentum. The vanishing of a coefficient of stability then requires

$$F_i \equiv \frac{1}{3} L_1 S_1 - \frac{1}{2n+1} L_i S_i = 0 \quad (\lambda = 0), \tag{1}$$

and for the spheroids we have, since $a = b$,

$$L_i = (1 + \tau^2)^{\frac{1}{2}p} D^{p+n} (1 + \tau^2)^n \quad (p = 0, 1, \ldots, n),$$

where
$$\tau^2 = (c^2 + \lambda)/(a^2 - c^2).$$

Now equation (1) involves only a and c (in effect their ratio) since λ is put zero, but if regarded as an equation for a^2 it is clear that λ can be retained and the equation equally well be thought of as one in $\lambda + a^2$. In the following discussion it is found more convenient to proceed in this way; that is, λ is not put zero but instead $\lambda + a^2$ is treated as the unknown in the equation.

It is to be noticed that when $a = b$, although the total number of ellipsoidal harmonics of order n is still $2n+1$, as in the general case, there are only $n+1$ different L-functions (and M-functions). To each of these, however, there correspond two different N-functions, namely $\cos p\phi$, $\sin p\phi$, except for $p = 0$, when there is only one, namely $N = 1$. This makes $2n+1$ harmonics in all, but only $n+1$ different coefficients of stability, since these depend only on the L- and S-functions. Thus of the $2n+1$ coefficients of stability, $2n$ are equal in pairs, and if any one of these (for which $p \neq 0$) should vanish, it means two coefficients vanishing simultaneously.

The spheroid common to the Maclaurin and Jacobi series

We first show that, when $a = b$, the equation

$$F_2 \equiv \tfrac{1}{3} L_1 S_1 - \tfrac{1}{5} L_2 S_2 = 0 \quad (\lambda = 0), \tag{2}$$

wherein the L_2 concerned (of the three such second-order Lamé functions) corresponds to $p = 2, n = 2$, has a unique root and that it corresponds to the point on the Maclaurin series at which the Jacobi series branches off. That is, it corresponds to the point of birfurcation on these series already established when certain restricted second-order displacements only are permitted (pp. 45 *et seq.*).

In the present case we have

$$L_2 = (1+\tau^2)D^4(1+\tau^2)^2$$

$$= a^2+\lambda, \quad \text{apart from positive constant factors.}$$

Whence $\qquad S_2 = (a^2+\lambda)\int_\lambda^\infty \dfrac{5d\lambda}{2(a^2+\lambda)^2\sqrt{[(a^2+\lambda)(b^2+\lambda)(c^2+\lambda)]}}.$

Also we have $\qquad\qquad L_1 = \sqrt{(c^2+\lambda)},$

$$S_1 = \sqrt{(c^2+\lambda)}\int_\lambda^\infty \dfrac{3d\lambda}{2(c^2+\lambda)\sqrt{[(a^2+\lambda)(b^2+\lambda)(c^2+\lambda)]}}.$$

Substituting these expressions in the coefficient of stability and putting $\lambda = 0$, equation (2), written at length, is

$$c^2\int_0^\infty \dfrac{d\lambda}{(c^2+\lambda)\sqrt{[(a^2+\lambda)(b^2+\lambda)(c^2+\lambda)]}} - a^4\int_0^\infty \dfrac{d\lambda}{(a^2+\lambda)^2\sqrt{[(a^2+\lambda)(b^2+\lambda)(c^2+\lambda)]}} = 0,$$

and if we now put $a = b$ in this it becomes

$$c^2\int_0^\infty \dfrac{d\lambda}{(a^2+\lambda)(c^2+\lambda)\sqrt{(c^2+\lambda)}} - a^4\int_0^\infty \dfrac{d\lambda}{(a^2+\lambda)^3\sqrt{(c^2+\lambda)}} = 0, \qquad (3)$$

and this is identical with the condition for a Jacobi ellipsoid with $a = b$ put in it (p. 40). This establishes therefore not only that (2) has a root but that it is the only root, when $a = b$, for there is only one spheroidal member of the Jacobi series.

The equation $F_i = 0$

(It is particularly to be noticed that the arguments of the present section refer to the general case $a > b > c$ and therefore apply equally to the Maclaurin and Jacobi figures.)

If τ^2 is used as the variable, the spherical form $a = b = c$ corresponds to $\dfrac{\lambda+c^2}{a^2-c^2} = \infty$, that is, to $\tau^2 = \infty$. Again, the infinitely flattened disc spheroid $a = b = \infty$, $c = 0$ corresponds (since $\lambda = 0$) to $\lambda+c^2 = 0$, that is, to $\tau^2 = 0$. Hence if we regard τ^2 as gradually decreasing from ∞ to 0, the corresponding equilibrium configuration will describe the Maclaurin series from the spherical form to the infinite circular disc-spheroid of vanishing thickness.

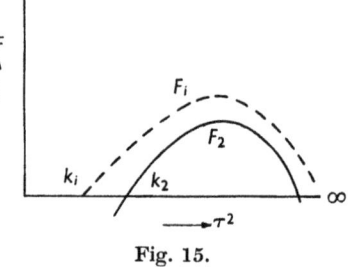

Fig. 15.

Consider then the graph of the function $F_2(\tau^2)$ over this range. Near $\tau^2 = \infty$, its value is (p. 80) approximately $(\tfrac{1}{3}-\tfrac{1}{5})\lambda^{-\frac{1}{2}}$ and the function is accordingly positive and decreasing for large λ. Let us denote by $\tau^2 = k_2$ the unique finite root of the equation. Then the graph of F_2 must be of the form shown in Fig. 15. That is, it must be positive at $\tau^2 = \infty$, begin by increasing as τ^2 decreases, and eventually cross the axis at $\tau^2 = k_2$. (There might, of course, be more than one maximum between ∞ and k_2, but the point is not relevant to the argument to be given.)

THE EQUATION $F_i = 0$

Now it will certainly establish that F_2 is the first coefficient of stability to vanish if it can be shown that for $\tau^2 \geqslant k_2$ any other coefficient of stability F_i exceeds F_2. (This result, which can in fact be established, is clearly a sufficient condition though not a necessary one.) That is, it will suffice if it can be shown that

$$\frac{1}{3} L_1 S_1 - \frac{1}{2n+1} L_i S_i > \frac{1}{3} L_1 S_1 - \frac{1}{5} L_2 S_2 \qquad \text{for } \tau^2 \geqslant k_2;$$

that is, if
$$\frac{1}{5} L_2 S_2 - \frac{1}{2n+1} L_i S_i > 0 \qquad \text{for } \tau^2 \geqslant k_2.$$

It will be proved that this inequality holds strongly for all $\tau^2 > 0$, that is, for all $\lambda + a^2 > a^2 - c^2$, and this ensures that it holds for $\tau^2 \geqslant k_2$ since this value corresponds to the first Jacobi figure and must therefore be a positive quantity.

To establish this we begin by considering the more general equation in $\lambda + a^2$,

$$F \equiv \frac{1}{2m+1} L_k S_k - \frac{1}{2n+1} L_i S_i = 0, \qquad (4)$$

where m denotes the order of the Lamé function L_k. Let us examine the question of what roots, if any, this may have for $\tau^2 > 0$.

We notice first that since $LS \to \lambda^{-\frac{1}{2}}$ as $\lambda \to \infty$ for any order n, then $F \to 0$ as $\tau^2 \to \infty$. Also, it has been shown (p. 70) that all the roots of the rational part of the equation
$$L(\lambda + a^2) = 0$$

lie between 0 and $a^2 - c^2$, and that there is a zero at $a^2 - c^2$ only if L actually contains the radical $\sqrt{(\lambda + c^2)}$ as factor. Hence, if we consider only values of $\lambda + a^2$ essentially greater than $a^2 - c^2$ (that is, exclude $\lambda + a^2 = a^2 - c^2$), the roots of $F = 0$ are the same as of $F/K_k^2 = 0$; that is, the same as of the equation

$$\frac{1}{2m+1} \frac{S_k}{L_k} - \frac{1}{2n+1} \cdot \frac{S_i}{L_i} \cdot \frac{L_i^2}{L_k^2} = 0. \qquad (5)$$

If now this equation is differentiated with regard to λ, we shall get an equation whose roots separate those of (5). The derived equation is

$$\frac{d}{d\lambda} \left\{ \frac{1}{2m+1} \cdot \frac{S_k}{L_k} \right\} - \frac{L_i^2}{L_k^2} \cdot \frac{d}{d\lambda} \left\{ \frac{1}{2n+1} \cdot \frac{S_i}{L_i} \right\} - \frac{1}{2n+1} \cdot \frac{S_i}{L_i} \cdot \frac{d}{d\lambda} \left\{ \frac{L_i^2}{L_k^2} \right\} = 0. \qquad (6)$$

But by the definition of the S-functions

$$\frac{S}{L} = \int_\lambda^\infty \frac{(2n+1) d\lambda}{2L^2 A} \qquad \text{where } A^2 = (\lambda + a^2)(\lambda + b^2)(\lambda + c^2),$$

and hence
$$\frac{d}{d\lambda} \left\{ \frac{1}{2n+1} \cdot \frac{S}{L} \right\} = -\frac{1}{L^2} \cdot \frac{1}{2A}.$$

Using this result, the first two terms in (6) disappear, and the derived equation is simply
$$\frac{S_i}{L_i} \frac{d}{d\lambda} \left\{ \frac{L_i^2}{L_k^2} \right\} = 0, \qquad (7)$$

and its roots separate those of (5) or (4).

Since L_i and therefore also S_i are essentially positive over the range

$$a^2 - c^2 < \lambda + a^2 < \infty,$$

the factor S_i/L_i cannot vanish, and (7) simply expresses the condition for L_i/L_k to have a critical value in this range. Hence the values at which this occurs separate the roots of (4).

Differentiation with regard to λ may be replaced by differentiation with regard to u (p. 79) since $\dfrac{du}{d\lambda}$ cannot vanish for $\lambda + a^2 > a^2 - c^2$, so that (7) becomes

$$L_k \frac{dL_i}{du} - L_i \frac{dL_k}{du} = 0. \tag{8}$$

A further differentiation with regard to u will now provide an equation whose roots separate those of (8), namely

$$L_k \frac{d^2 L_i}{du^2} - L_i \frac{d^2 L_k}{du^2} = 0. \tag{9}$$

But Lamé's equation, with u as independent variable, for L_i and L_k gives

$$\frac{d^2 L_i}{du^2} = \{n(n+1)\lambda - K_i\} L_i,$$

$$\frac{d^2 L_k}{du^2} = \{m(m+1)\lambda - K_k\} L_k,$$

and inserting these in (9) and omitting the factor $L_i L_k$, which cannot vanish for $\lambda + a^2 > a^2 - c^2$, we arrive finally at the simple linear equation

$$\{n(n+1) - m(m+1)\}\lambda = K_i - K_k. \tag{10}$$

Since this is of the first degree it obviously can have at most *one* root exceeding $a^2 - c^2$. Hence (8) can have at most *two* such roots, and (4) can have at most *three* such roots, that is, values of $\lambda + a^2$ exceeding $a^2 - c^2$.

We notice also that if ∞ turns out to be a root of (4), as in fact will be shown to be so, it would count as one of the *three at most* whose possible existence is demonstrated by the above argument. This would then mean that (4) could have *at most two* finite roots $> a^2 - c^2$. We make important use of this result in case (ii) below.

Having established this for the general equation $F = 0$ we return to the original form of the equation (1), viz. $F_i = 0$. The discussion requires consideration of three cases, as follows:

(i) *Suppose L_i contains $\sqrt{(\lambda + c^2)}$ as factor.*

We may then write $\qquad L_i = \sqrt{(\lambda + c^2)}\phi(\lambda)f_i(\lambda),$

where $\qquad \phi(\lambda) = 1, \quad \sqrt{(\lambda + a^2)}, \quad \sqrt{(\lambda + b^2)}, \quad$ or $\quad \sqrt{[(\lambda + a^2)(\lambda + b^2)]},$

and $f_i(\lambda)$ is a rational polynomial in λ in which the highest power occurring has positive coefficient.

Now it has been shown that $f_i = 0$, as an equation in $\lambda + a^2$, has all its roots real and lying between 0 and $a^2 - c^2$. Hence $\dfrac{df_i}{d\lambda} = 0$ also has all its roots real and in this same interval, and hence for $\lambda + a^2$ increasing from $a^2 - c^2$ to ∞, $\dfrac{df_i}{d\lambda}$ has constant sign, and therefore f_i is always increasing, since $f_i(\infty) = +\infty$. The same also holds for both the radical factors $\sqrt{(\lambda + a^2)}$ and $\sqrt{(\lambda + b^2)}$, and hence in the wide sense for any possible $\phi(\lambda)$.

Hence in the present case

$$L_i/L_k = L_i/L_1 = \phi(\lambda)f_i(\lambda)$$

is always increasing with λ in the range $a^2 - c^2 < \lambda + a^2 < \infty$, which means that

$$\frac{d}{d\lambda}(L_i^2/L_k^2) > 0,$$

and hence, by (5) and (6), that

$$\frac{d}{d\lambda}(F_i/L_k^2) < 0.$$

Accordingly F_i/L_k^2 is always decreasing as λ increases, and since it $\to +0$ as $\lambda \to \infty$, it must be always positive for finite $\lambda + a^2$ exceeding $a^2 - c^2$. We thus have the following result (for $\tau^2 > 0$):

If L_i contains $\sqrt{(\lambda + c^2)}$ as factor the equation $F_i = 0$ has no root.

We notice that in the case $a = b$, the radical $\sqrt{(\lambda + c^2)}$ being a factor means that τ is a factor of L when expressed in the form

$$L = (1 + \tau^2)^{\frac{1}{2}p} D^{p+n}(1 + \tau^2)^n.$$

But clearly L has τ as factor whenever $p + n$ is odd, and hence in the case of spheroids for there to be a possibility of $F_i = 0$ possessing a root, the L-function must be such that $p + n$ *is even*.

(ii) *Suppose L_i does not contain $\sqrt{(\lambda + c^2)}$ as factor.*

Assuming $n \geqslant 2$, we have for λ large

$$\frac{1}{3}L_1 S_1 - \frac{1}{2n+1}L_i S_i = \left(\frac{1}{3} - \frac{1}{2n+1}\right)\lambda^{-\frac{1}{2}} > 0 \quad \text{for} \quad \lambda < \infty$$

$$\text{and} \quad = 0 \quad \text{for} \quad \lambda = \infty.$$

For $\lambda + a^2 = a^2 - c^2$, since $L_1 = \sqrt{(\lambda + c^2)}$ we have $L_1 S_1 = 0$, whereas, on the present hypothesis, $L_i S_i$ does not vanish for $\lambda + c^2 = 0$. But since $L_i = 0$ has no root $\geqslant a^2 - c^2$, the product $L_i S_i$ has the same sign for all values of $\lambda + a^2 > a^2 - c^2$, and near infinity this sign is plus. Hence for $\lambda + a^2 = a^2 - c^2$ we have

$$\frac{1}{3}L_1 S_1 - \frac{1}{2n+1}L_i S_i < 0.$$

The function F_i therefore changes sign for some finite $\lambda + a^2$ exceeding $a^2 - c^2$, and the number of roots *between* $a^2 - c^2$ and ∞ (not counting ∞ as a root) must therefore be odd. Infinity itself is also a root. But it was shown above that this equation can have at most *two* finite roots, and since now it is necessary that the number must be odd, it follows that *the actual number of finite roots must be just one.*

Hence every L_i for spheroids for which $p + n$ is even gives rise to an equation $F_i = 0$ possessing a single root $\lambda + a^2 > a^2 - c^2$. That is, for some positive finite value of $\lambda + c^2$ (before the infinite disc spheroid is reached) the coefficient of stability corresponding to every L_i for which $p + n$ is even will vanish and change sign. Thus secular instability will enter at some stage for each of the corresponding harmonic displacements.

It will be shown later that the particular L_i for which the spheroid first becomes unstable is $L_2 = 1 + \tau^2$; that is, the second order harmonic corresponding to $n = 2$, $p = 2$. But before proceeding to show this we consider the remaining possible form for L_i.

(iii) *Suppose $n = 1$ so that $L_i = \sqrt{(\lambda + a^2)}$ or $\sqrt{(\lambda + b^2)}$.*

It has already been shown (p. 90) that the displacements corresponding to first-order harmonics consist of (small) translations of the ellipsoid parallel to the co-ordinate axes. For the third of these, viz. $\sqrt{(\lambda + c^2)}$, the coefficient of stability is zero identically, and the equilibrium therefore neutral for the corresponding displacement, as already explained.

Where the first two are concerned, it is clear since $a > b > c$ that both the ratios

$$\frac{\sqrt{(\lambda + a^2)}}{\sqrt{(\lambda + c^2)}} \quad \text{and} \quad \frac{\sqrt{(\lambda + b^2)}}{\sqrt{(\lambda + c^2)}}$$

are always decreasing as $\lambda + a^2$ increases from $a^2 - c^2$ to ∞, and their final values are both unity. Hence they are always positive, and equation (1) accordingly has no finite root $> a^2 - c^2$, and it follows that the corresponding coefficients of stability are always negative.

But, as already mentioned in Chapter VI, such deformations are ruled out since they involve a displacement of the centre of gravity off the axis of rotation, which would indicate a continued general translatory motion of the whole mass within the adopted frame of reference, whereas it is necessary to consider only motion relative to axes fixed at the centre of gravity.

As far as the Maclaurin spheroids are concerned, (i), (ii), and (iii) above cover all possible cases. Before proceeding it may be convenient to summarize the results of these foregoing sections on the number of roots of the equation $F_i(\lambda + a^2) = 0$ lying between $a^2 - c^2$ and ∞.

(i) *If L_i is divisible by $\sqrt{(\lambda + c^2)}$.*   0 root; $\dfrac{1}{3} L_1 S_1 - \dfrac{1}{2n+1} L_i S_i > 0$ always.

(ii) *If L_i is not divisible by $\sqrt{(\lambda + c^2)}$ and $n \geqslant 2$.*   1 finite root;

$\dfrac{1}{3} L_1 S_1 - \dfrac{1}{2n+1} L_i S_i$ at first $(\tau^2 = \infty)$ positive, but becoming negative beyond some finite τ^2.

(iii) *If L_i is not divisible by $\sqrt{(\lambda + c^2)}$ and $n = 1$.* 0 root; and the coefficient of stability always negative. But this type of deformation is ruled out as it displaces the centre of gravity off the axis of rotation.

It may again be mentioned that these results hold generally for all $a \geqslant b > c$, and therefore apply equally to the discussions of the Maclaurin series and the Jacobi series. We proceed now to establish the following result for the Maclaurin series.

The first coefficient of stability to vanish is that corresponding to L_2 ($p = 2$).

As the mass gradually flattens from the spherical form and describes the Maclaurin series, the first coefficient of stability to vanish will be shown to be that corresponding to

$$L_2 = (1 + \tau^2) D^4 (1 + \tau^2)^2, \qquad \text{i.e. } p = 2, n = 2$$
$$= 1 + \tau^2, \qquad \text{apart from a constant factor.}$$

As far as L_i is concerned, it is only necessary to consider $p + n$ even, since it has been proved that there can be no root otherwise. We wish, therefore, to show that the root $\tau^2 = k_2$ (corresponding to the spheroidal member of the Jacobi series) of the equation

$$\tfrac{1}{3} L_1 S_1 - \tfrac{1}{5} L_2 S_2 = 0 \qquad (p = 2)$$

is greater than (since τ^2 decreases from ∞ as the spheroid flattens) the root $\tau^2 = k_i$ of

$$\frac{1}{3} L_1 S_1 - \frac{1}{2n+1} L_i S_i = 0.$$

For this purpose, as has been seen, it will certainly suffice if it can be proved that

$$\frac{1}{3} L_1 S_1 - \frac{1}{2n+1} L_i S_i > \frac{1}{3} L_1 S_1 - \frac{1}{5} L_2 S_2$$

or that

$$\frac{1}{5} L_2 S_2 - \frac{1}{2n+1} L_i S_i > 0,$$

and this has already been seen to be equivalent to proving that L_i/L_2 is always increasing with τ^2.

To prove this it is necessary to consider three separate forms for L_i covering all cases. It is important to notice that we are again able to establish sufficient conditions; that is, the above inequality in fact holds strongly and we are able to take advantage of this in the proof.

Case (a). Suppose $p \geqslant 2$. Then

$$L_i/L_2 = (1 + \tau^2)^{\frac{1}{2}(p-2)} D^{p+n} (1 + \tau^2)^n, \qquad p + n \text{ is even.}$$

In this expression, the first factor $(1 + \tau^2)^{\frac{1}{2}(p-2)}$ either reduces to 1 or else is always increasing with τ^2, according as $p = 2$ or $p > 2$. Again, since $(1 + \tau^2)^n$ is a polynomial in τ^2 with all its coefficients positive, so also must be $D^{p+n}(1 + \tau^2)^n$, and it is therefore always an increasing function of τ^2. Hence L_i/L_2 is always increasing, and accordingly it follows that for L_i of the present type the coefficient of stability corresponding to L_2 must vanish first.

7

Case (*b*). Suppose $p = 0$; then $n = 2j$, where j is an integer $\geqslant 1$ (since $p+n$ is necessarily even).

We now have
$$L_i/L_2 = \{D^{2j}(1+\tau^2)^{2j}\}/(1+\tau^2).$$

(This form includes the remaining Lamé function $L = 1+3\tau^2$ of order 2, given by $D^2(1+\tau^2)^2$. The second function $L = \tau\sqrt{(1+\tau^2)}$ contains τ as factor and its corresponding coefficient of stability cannot vanish.)

In this case L_i/L_2 is always positive since both the numerator and denominator are always positive. If therefore it can be shown that its differential coefficient is always positive it would follow that it is always increasing with τ. The logarithmic differential coefficient is

$$\frac{D^{2j+1}(\tau^2+1)^{2j}}{D^{2j}(\tau+1)^{2j}} - \frac{2\tau}{\tau^2+1},$$

and what we wish to show is equivalent to proving that

$$E(\tau) \equiv (\tau^2+1)\,D^{2j+1}(\tau^2+1)^{2j} - 2\tau\,D^{2j}(\tau^2+1)^{2j} > 0.$$

Now
$$(\tau^2+1)^{2j} = \sum_k \frac{2j!}{k!\,(2j-k)!}\,\tau^{2k} \qquad (k = 0, 1, \ldots, 2j),$$

so that

$$D^{2j}(\tau^2+1)^{2j} = \sum_k \frac{2j!\,2k!}{k!\,(2j-k)!\,(2k-2j)!}\,\tau^{2k-2j} \quad (k = j, j+1, \ldots, 2j)$$

$$= \sum_k A_k\tau^{2k-2j}, \quad \text{say},$$

and it is seen that all the coefficients A_k are essentially positive. We thus have

$$E(\tau) = (\tau^2+1)\Sigma A_k(2k-2j)\tau^{2k-2j-1} - 2\tau\Sigma A_k\tau^{2k-2j},$$

and it will be shown that this polynomial in τ has all its coefficients positive. The coefficient of $\tau^{2k-2j+1}$ is found to be

$$(2k-2j-2)\,A_k + (2k-2j+2)\,A_{k+1}.$$

Since $A_k > 0$ and $k \geqslant j$, the only one of these coefficients that might be negative is that for which $k = j$, and this one is $-2A_j + 2A_{j+1}$, since for $k > j$ all the factors multiplying A_k and A_{k+1} are positive. But this one coefficient will also be positive if $A_{j+1} > A_j$. That is if

$$\frac{2j!\,(2j+2)!}{(j+1)!\,(j-1)!\,2!} > \frac{2j!\,2j!}{j!\,j!},$$

that is, if
$$(2j+2)\,(2j+1) > 2(j+1)/j,$$

or if
$$j(2j+1) > 1,$$

and since $j \geqslant 1$, this is certainly satisfied.

Thus all the coefficients in $E(\tau)$ are positive, and therefore L_i/L_2 is positive and always increasing with τ^2. Hence the coefficient of stability corresponding to $L_2 = 1+\tau^2$ vanishes before any of those corresponding to L_i of the present type as the mass gradually evolves along the spheroidal series from the spherical form.

Case (c). Lastly, we have $p = 1$, $n = 2j + 1$ (since $p + n$ is even). It is to be noticed that $j \geqslant 1$ since $j = 0$ would correspond to a first-order harmonic, and these are already excluded.

In the present case

$$L_i = (1 + \tau^2)^{\frac{1}{2}} D^{2j+2} (1 + \tau^2)^{2j+1},$$

and hence
$$L_i / L_2 = \{ D^{2j+2} (1 + \tau^2)^{2j+1} \} / (1 + \tau^2)^{\frac{1}{2}}.$$

This will be always a positive increasing function of τ if its logarithmic differential coefficient is always positive, that is, if

$$\frac{D^{2j+3} (\tau^2 + 1)^{2j+1}}{D^{2j+2} (\tau^2 + 1)^{2j+1}} - \frac{\tau}{\tau^2 + 1}$$

is always positive, and this will be so if

$$E(\tau) \equiv (\tau^2 + 1) D^{2j+3} (\tau^2 + 1)^{2j+1} - \tau D^{2j+2} (\tau^2 + 1)^{2j+1} > 0.$$

Proceeding in a similar way as in case (*b*), we have

$$D^{2j+2} (\tau^2 + 1)^{2j+1} = \sum_k \frac{(2j+1)! \, 2k!}{k! \, (2j-k+1)! \, (2k-2j-2)!} \tau^{2k-2j-2}$$

$$= \sum_k B_k \tau^{2k-2j-2}, \quad \text{say}, \quad (k = j+1, j+2, ..., 2j+1),$$

and here all the coefficients B_k are positive. Thus we have

$$E(\tau) = (\tau^2 + 1) \sum (2k - 2j - 2) B_k \tau^{2k-2j-3} - \tau \sum B_k \tau^{2k-2j-2},$$

and in this the coefficient of $\tau^{2k-2j-1}$ is

$$(2k - 2j - 3) B_k + (2k - 2j) B_{k+1}.$$

This might possibly be negative if $2k < 2j + 3$, that is, if $k = j + 1$, in which event it becomes $2B_{j+2} - B_{j+1}$. But even this will be positive if

$$\frac{2 \cdot (2j+1)! \, (2j+4)!}{(j+2)! \, (j-1)! \, 2!} > \frac{(2j+1)! \, (2j+2)!}{(j+1)! \, j!}$$

that is, if
$$\frac{(2j+4)(2j+3)}{j+2} > \frac{1}{j},$$

or if
$$2j(2j+3) > 1,$$

and this is certainly satisfied even for $j = 1$.

Accordingly it follows that the coefficient of stability corresponding to $L_2 = 1 + \tau^2$ vanishes before that corresponding to any L_i of the present form.

Taking these three cases together we may therefore conclude that as τ^2 decreases from ∞, the first coefficient of stability to vanish is that associated with the second-order Lamé function $L_2 = a^2 + \lambda$. Thus *secular stability* of the Maclaurin series is lost at this point. It is to be remembered, however, that the present result has been reached from consideration of the function $W - \frac{1}{2} \omega^2 I$, and this corresponds to stability relative to a frame of reference rotating with constant angular velocity. Since the harmonics associated with this L_2 are the second-order polynomials $x^2 - y^2$ and xy, it is possible that ω is changed to the first order of small quantities as a result of the corresponding surface deformations. Accordingly it is necessary to establish that the instability arrived at here is a true instability (see p. 26).

As far as xy is concerned, it is readily shown that a deformation ξ/ϖ involving the surface harmonic MN to which xy reduces on the ellipsoid will not affect the moment of inertia I to the first order. For it has been seen (p. 83) that the first-order change in I is

$$\delta_1 I = \int_E (x^2 + y^2)\, \xi dS$$

and on the surface $x^2 + y^2$ can be expressed in terms of solid harmonics in the form

$$h(x^2 - y^2) + k(y^2 - z^2) + l$$

where h, k, and l are constants. Accordingly, since xy is a harmonic independent of this, the surface integral giving $\delta_1 I$ vanishes. Thus I is changed only to second order and the instability indicated by the vanishing of the coefficient of stability associated with xy must be a true one.

On the other hand the surface deformation corresponding to $x^2 - y^2$ will produce in $\delta_1 I$ a term of the form

$$\int \varpi L_2^2 M_2^2 dS$$

and the moment of inertia is therefore changed to the first order. But the surface harmonics corresponding to $x^2 - y^2$ and $y^2 - z^2$ are precisely those involved in the restricted ellipsoidal displacements considered in Chapter IV (p. 45). It is easily verified that the normal displacement ξ required to deform the standard ellipsoid

$$\frac{x^2}{a^2} + \frac{y^2}{b^2} + \frac{z^2}{c^2} = 1$$

into a slightly different one with common principal axes, say

$$\frac{x^2}{(a + \delta a)^2} + \frac{y^2}{(b + \delta b)^2} + \frac{z^2}{(c + \delta c)^2} = 1 \quad \text{with} \quad \delta(abc) = 0$$

when made harmonic, is given by an expression of the form

$$\xi/\varpi = p(x^2 - y^2) + q(y^2 - z^2) + r,$$

where p, q, r are small constants vanishing if $\delta a, \delta b, \delta c$ all vanish. Accordingly the argument given in Chapter IV, which is based on the function $W + H^2/2I$ appropriate to systems with changing angular velocity, can be adduced to show that the instability corresponding to $x^2 - y^2$ is also a true instability.

Thus it may finally be concluded that the Maclaurin spheroid is secularly stable, and therefore also ordinarily stable, for all deformations (representable by ellipsoidal surface harmonic expansions) provided its eccentricity of meridian section is less than that of the spheroid of bifurcation, that is, for $e < 0.8127$. For values greater than this the spheroid is secularly unstable.

The question of its ordinary stability, which cannot of course be settled from the present considerations, has been studied by Cartan, who has shown that the spheroidal form remains ordinarily stable for all possible small oscillations (representable by ellipsoidal surface harmonics) provided the eccentricity of section is less than 0·9529.

From a practical standpoint, however, the secularly unstable spheroids are of little interest, for if we are regarding the mass as undergoing evolution owing to a gradual increase of angular momentum, then when the value exceeds that corresponding to the first point of bifurcation, the mass would proceed to evolve along the Jacobi series, which (as will be shown in the next chapter) begins by being thoroughly stable. The departure from axial symmetry thereby involved may be regarded as brought about by some unspecified small external disturbance which can always be regarded as operative in actual physical systems.

The Maclaurin spheroids beyond the first form of bifurcation

If the series of spheroids is followed beyond the point at which the Jacobi series bifurcates, and regarded as subjected at any stage only to harmonic deformations of a given order n, then it can be proved that the first coefficient of stability to vanish corresponds always to the highest value of the index p, that is to $p = n$. In other words, the first coefficient of stability to vanish is that corresponding to the Lamé function

$$L = (1 + \tau^2)^{\frac{1}{2}n}.$$

It may further be proved that the coefficients of stability corresponding to $p = n - 2$, $p = n - 4$, ... vanish successively as the series continues to be described in the direction of increasing eccentricity and angular momentum. But it does *not* hold that all such instabilities of a given order n are reached *before* the first of those entering from harmonics of the higher orders $n + 1$, $n + 2$, ...

Since, however, all such configurations would already have become secularly unstable for displacements corresponding to $L(n = 2, p = 2)$ they are of no physical application and could not come into existence through any natural evolution of a liquid mass. If a system possessed an amount of angular momentum appropriate to any such spheroidal form, it would in fact settle down through the action of internal friction to the corresponding equilibrium configuration on the Jacobi series provided such configuration (with this angular momentum) is itself secularly stable. Accordingly we proceed next to consider the secular stability of the ellipsoidal forms.

Chapter VIII

THE SECULAR STABILITY OF THE JACOBI ELLIPSOIDS

The object of the discussion of the present chapter is to discover at what stages on the Jacobi series forms of bifurcation are reached. The main result from a practical standpoint is that the first coefficient of stability to vanish is that corresponding to a certain third-order harmonic.

We have now the general case $a > b > c$. As usual we put

$$\frac{\lambda + c^2}{a^2 - c^2} = \tau^2, \quad \text{so that} \quad \frac{\lambda + a^2}{a^2 - c^2} = \tau^2 + 1.$$

It is found convenient to represent description of the Jacobi series, in the direction of increasing angular momentum, by making $a^2 - b^2$ vary from 0 to $a^2 - c^2$. The value $a^2 - b^2 = 0$ clearly corresponds to the initial spheroidal form, while $a^2 - b^2 = a^2 - c^2$, that is $b = c$, corresponds to the infinitely elongated final Jacobi form $a = \infty, b = c = 0$ (Table II, p. 41). To each value of $a^2 - b^2$ in this range there will correspond a certain Jacobi figure, and hence a certain value of τ (since λ is always zero). That is, for given $\dfrac{a^2 - b^2}{a^2 - c^2}$ between 0 and 1 there is a unique Jacobi ellipsoid, and therefore definite values of the ratios $a : b : c$, or of a, b and c themselves if we suppose $abc = 1$ always. In terms of τ the axes of this ellipsoid will be

$$a = \sqrt{(\tau^2 + 1)} \cdot \sqrt{(a^2 - c^2)},$$

$$b = \sqrt{\left(\tau^2 + 1 - \frac{a^2 - b^2}{a^2 - c^2} \right)} \cdot \sqrt{(a^2 - c^2)},$$

$$c = \tau \sqrt{(a^2 - c^2)},$$

so that

$$a : b : c = \sqrt{(\tau^2 + 1)} : \sqrt{\left(\tau^2 + 1 - \frac{a^2 - b^2}{a^2 - c^2} \right)} : \tau.$$

The condition for a Jacobi ellipsoid

This condition has already been found from first principles in Chapter IV (p. 38) but it can be derived quite simply in a form more suitable for our purpose in the following way.

Suppose we have an equilibrium Jacobi figure E, and it is given a rigid body rotation through an infinitesimal angle θ round the z-axis, so that referred to the axes of E it is a new ellipsoid E' of the same size and shape as E. If the normal distance of points of E' from E is denoted by ξ, it is easily found that

$$\xi = \theta \left(\frac{1}{b^2} - \frac{1}{c^2} \right) \varpi M_2 N_2,$$

where $M_2 N_2$ is that one of the five second-order surface harmonics given by

$$M_2 = \sqrt{[(\mu + a^2)(\mu + b^2)]}, \quad N_2 = \sqrt{[(\nu + a^2)(\nu + b^2)]}.$$

For a displacement of the present kind it is evident that equilibrium must be neutral, for E' is simply a Jacobi form referred to different axes. Hence the corresponding coefficient of stability must vanish. That is

$$\tfrac{1}{3} L_1 S_1 - \tfrac{1}{5} L_2 S_2 = 0 \quad (\lambda = 0), \tag{1}$$

and this must accordingly be the condition for a Jacobi ellipsoid. It is easily verified that it is equivalent to the condition already found in Chapter IV.

It then follows from Table II that for a given value of $\dfrac{a^2 - b^2}{a^2 - c^2}$ between 0 and 1, the equation has just one real solution $\tau = j_2$, say.

Further, if $b \to c$ for a configuration on the Jacobi series the axes become in the ratios $a : b : c = 1 : 0 : 0$, and hence $\tau = 0$. Accordingly as $\dfrac{a^2 - b^2}{a^2 - c^2} \to 1$, the root $j_2 \to 0$.

Condition for a point of bifurcation

For this it is necessary that another coefficient of stability vanishes. That is, in addition to (1), we must have

$$\frac{1}{3} L_1 S_1 - \frac{1}{2n+1} L_i S_i = 0. \tag{2}$$

Hence for a point of bifurcation on a Jacobi figure we require simultaneously

$$\frac{1}{3} L_1 S_1 = \frac{1}{5} L_2 S_2 = \frac{1}{2n+1} L_i S_i \qquad (\lambda = 0), \tag{3}$$

where $L_1 = \sqrt{(\lambda + c^2)}$, and $L_2 = \sqrt{[(\lambda + a^2)(\lambda + b^2)]}$.

Now it has already been proved (p. 94) that equation (2) cannot have any real root if L_i is divisible by $\sqrt{(\lambda + c^2)}$, and that if it is not so divisible then it has *one* real finite root. Let us suppose that when L_i is not divisible by $\sqrt{(\lambda + c^2)}$ the root of (2) is denoted by $\tau = k_i$. Then for both of equations (3) to be satisfied simultaneously requires

$$j_2 = k_i.$$

It will now be shown that for a given order n, there must be at least one L_i for which a solution of (3) exists. To do this let us consider first the situation at the two ends of the range, viz. $a = b$, and $b = c$.

(i) $a = b$. This condition alone means that the ellipsoid is always on the Maclaurin series; $\tau = \infty$ corresponds to the sphere and $\tau = 0$ to the infinite disc. The root $\tau = j_2$ of (1), taken in conjunction with $a = b$, gives the first member of the Jacobi series, and the vanishing of any other coefficient of stability (on the spheroidal series) has been shown to occur for a value of τ smaller than j_2, that is, at a more flattened form. Accordingly, at this end of the range we have

$$j_2 > k_i \quad \text{for } a = b.$$

(ii) $b = c$. It has been seen (p. 63) that in this case the function L_i is given by

$$L_i = (t^2 - 1)^{\frac{1}{2}p} D^{p+n}(t^2 - 1)^n \quad (p = 0, 1, 2, ..., n),$$

where $t^2 = (a^2 + \lambda)/(a^2 - c^2)$. Also we have

$$L_1 = \sqrt{(\lambda + c^2)} = \sqrt{(t^2 - 1)} = \tau.$$

Hence if $p \neq 0$, equation (2) is satisfied for $t = 1$, or $\tau = 0$, since then both L_1 and L_i vanish separately. Consequently in such cases the root k_i is zero, and therefore we have

$$j_2 = k_i, \quad (i = 2) \quad \text{for } b = c.$$

If, however, $p = 0$, we have

$$L_i = D^n(t^2 - 1)^n,$$

and $t = 1$ is no longer a solution of (2) since L_i does not vanish. Accordingly the solution now is not $\tau = 0$ but must be some finite positive value. Hence in this case we must have

$$k_i > j_2(= 0) \quad \text{for } b = c.$$

Hence for that particular function L_i of given order n that reduces when $b = c$ to the L-function for which $p = 0$, the equation $j_2 = k_i$ is satisfied at least once, and possibly an odd number of times, for some finite value, or values, of τ. Also, when $b = c$ there is only one L-function (for each given order n) for which $p = 0$; that is, only one of the $2n + 1$ general L-functions of given order reduces for $b = c$ to that particular L_i for which $p = 0$.

It is thus established, by combining the results of these two cases, that the coefficient of stability corresponding to this particular general L-function certainly vanishes for some intermediate $a : b : c$. It will be shown in due course that it is in fact the only one that can do so. To distinguish it from the remaining $2n$ coefficients of stability, which turn out to be always positive, it is termed 'the characteristic coefficient of stability of order n'.

The equation $\frac{1}{5} L_2 S_2 - \frac{1}{2n+1} L_i S_i = 0$ cannot have a root if L_i has $\sqrt{(\lambda + b^2)}$ as a factor

Let us now take the equation

$$\frac{1}{5} L_2 S_2 - \frac{1}{2n+1} L_i S_i = 0, \tag{4}$$

and in order to find the forms of L_i for which it *cannot* have a root let us consider in what cases L_i/L_2 is always increasing.

Omitting L-functions divisible by $\sqrt{(\lambda + c^2)}$ (for which it has already been seen the equation can have no root), the functions L_i can take four different possible forms, and written in factorized form these are as follows:

(i) $L_i = (\lambda + a^2 - \alpha_1)(\lambda + a^2 - \alpha_2) \dots (\lambda + a^2 - \alpha_m)$ for

(ii) $L_i = \sqrt{[(\lambda + a^2)(\lambda + b^2)]} \cdot (\lambda + a^2 - \alpha_1) \dots (\lambda + a^2 - \alpha_{m-1})$ $n = 2m$ $(m \geqslant 1)$

and

(iii) $L_i = \sqrt{(\lambda + a^2)} \cdot (\lambda + a^2 - \alpha_1) \dots (\lambda + a^2 - \alpha_m)$ for

(iv) $L_i = \sqrt{(\lambda + b^2)} \cdot (\lambda + a^2 - \alpha_1) \dots (\lambda + a^2 - \alpha_m)$ $n = 2m+1$ $(m \geqslant 1)$.

(It will be understood that in these expressions the roots are denoted conventionally by $\alpha_1, \alpha_2, \alpha_3, \dots$ in each of the forms without implying that the roots are the same in each expression.)

In each of the four expressions we may suppose the factors arranged in decreasing order of the α's, that is, so that

$$\alpha_1 > \alpha_2 > \alpha_3 > \dots.$$

Now it has been shown (p. 70) that for any L_i the roots in $\lambda + a^2$ are all real and positive and less than $a^2 - c^2$. That is,

$$0 < \alpha < a^2 - c^2 \quad \text{for every } \alpha.$$

Hence
$$\lambda + a^2 - \alpha > \lambda + c^2 \geqslant 0,$$

and therefore all the factors $\lambda + a^2 - \alpha$ will be positive increasing quantities as $\lambda + a^2$ increases from $a^2 - c^2$ to ∞.

We have also that for $\lambda + a^2$ increasing

$$\frac{\lambda + a^2 - \alpha}{\sqrt{(\lambda + a^2)}} = \sqrt{(\lambda + a^2)} - \frac{\alpha}{\sqrt{(\lambda + a^2)}}$$

$$= \text{increasing quantity} - \text{decreasing quantity}$$

$$= \text{increasing quantity for every } \alpha.$$

Also
$$\frac{\lambda + a^2 - \alpha}{\sqrt{(\lambda + b^2)}} = \sqrt{(\lambda + b^2)} - \frac{\alpha - (a^2 - b^2)}{\sqrt{(\lambda + b^2)}},$$

and this will certainly be always increasing provided $\alpha > a^2 - b^2$.

Now for the Jacobi ellipsoids $L_2 = \sqrt{[(\lambda + a^2)(\lambda + b^2)]}$, and hence the ratio L_i/L_2 can take four possible forms corresponding to the above, and these can be written:

(i) $\dfrac{L_i}{L_2} = \left\{ \dfrac{\lambda + a^2 - \alpha_1}{\sqrt{(\lambda + b^2)}} \right\} \left(\dfrac{\lambda + a^2 - \alpha_2}{\sqrt{(\lambda + a^2)}} \right) (\lambda + a^2 - \alpha_3) \quad \dots \quad (\lambda + a^2 - \alpha_m)$ n even

(ii) $\dfrac{L_i}{L_2} = \qquad\qquad\qquad\qquad\qquad (\lambda + a^2 - \alpha_1) \quad \dots \quad (\lambda + a^2 - \alpha_{m-1})$

and

(iii) $\dfrac{L_i}{L_2} = \left\{ \dfrac{\lambda + a^2 - \alpha_1}{\sqrt{(\lambda + b^2)}} \right\} (\lambda + a^2 - \alpha_2) \quad \dots \quad (\lambda + a^2 - \alpha_m)$ n odd.

(iv) $\dfrac{L_i}{L_2} = \left(\dfrac{\lambda + a^2 - \alpha_1}{\sqrt{(\lambda + a^2)}} \right) (\lambda + a^2 - \alpha_2) \quad \dots \quad (\lambda + a^2 - \alpha_m)$

In these four products all factors involved, *except for the first factor* {} *in each of* (i) *and* (iii), have already been shown to be always positive and increasing.

Hence in cases (ii) and (iv) the ratio L_i/L_2 is always increasing, and it is seen that these are precisely the cases in which L_i is divisible by $\sqrt{(\lambda+b^2)}$.

In cases (i) and (iii) L_i/L_2 will be always increasing provided $\alpha_1 > a^2 - b^2$. That is, (4) can only be satisfied if the greatest of the roots $\alpha_1, \alpha_2, \ldots$ is less than $a^2 - b^2$, and this means that *all* the roots of the rational part of L_i must be less than $a^2 - b^2$.

Hence we can conclude that if equation (4) is to have a root, L_i must not be divisible by $\sqrt{(\lambda+b^2)}$ and also all its zeros (of the rational part) must be less than $a^2 - b^2$. (It is not yet established that it does actually have a root, but only that it certainly cannot do so if these conditions are not satisfied by L_i.)

Only one coefficient of stability of given order *n* is capable of vanishing

To establish this it is necessary to consider the cases of *n* odd or even separately.

(i) *Suppose n even.* The *L*-functions can in the first instance take only one of the following four forms:

$$L_r, \quad \sqrt{[(\lambda+b^2)(\lambda+c^2)]}\,L_r, \quad \sqrt{[(\lambda+c^2)(\lambda+a^2)]}\,L_r, \quad \sqrt{[(\lambda+a^2)(\lambda+b^2)]}\,L_r,$$

where L_r denotes a rational polynomial in λ. But it was shown in Chapter VII that no coefficient of stability can vanish corresponding to an L_i that is divisible by $\sqrt{(\lambda+c^2)}$, and it has been shown in the foregoing section that it cannot vanish if L_i is divisible by $\sqrt{(\lambda+b^2)}$. Accordingly, for *n* even, the last three of the above forms are ruled out and only a coefficient of stability corresponding to a wholly rational L_i can possibly vanish. For these rational functions, $\kappa_1 = \kappa_2 = \kappa_3 = 0$ (see p. 57), and there are $\frac{1}{2}(n+2) = m+1$ such functions. But by Stieltjes' theorem, only *one* of these has all its zeros less than $a^2 - b^2$, and this is that one for which $r = m+1 = \frac{1}{2}n+1$ (in the notation of the proof of this theorem).

(ii) *Suppose n odd.* To begin with, the possible forms for L_i are now

$$\sqrt{(\lambda+a^2)}\,L_r, \quad \sqrt{(\lambda+b^2)}\,L_r, \quad \sqrt{(\lambda+c^2)}\,L_r, \quad \sqrt{[(\lambda+a^2)(\lambda+b^2)(\lambda+c^2)]}\,L_r,$$

where L_r denotes a rational polynomial, and again the last three forms are ruled out because of the presence of $\sqrt{(\lambda+b^2)}$, or $\sqrt{(\lambda+c^2)}$, or both, as a factor. There remains only the type $\sqrt{(\lambda+a^2)}\,L_r$, and since now $\kappa_1 = \frac{1}{2}$, $\kappa_2 = \kappa_3 = 0$, there are $\frac{1}{2}(n+1) = m+1$ of these in all. Again, however, only *one* of these has all its zeros less than $a^2 - b^2$, namely that for which $r = m+1 = \frac{1}{2}(n+1)$.

Hence in either event, *n* odd or even, only one coefficient of stability (of the $2n+1$ in all of a given order *n*) can vanish. This one has been termed the characteristic coefficient of stability of order *n*.

This does not of itself establish that the characteristic coefficient does vanish for some $a:b:c$, but the result now follows from the conclusion reached on p. 104, above. For it was shown that there is certainly one function L_i (of the $2n+1$ in all) for which $j_2 = k_i$ is satisfied, that is, for which the corresponding coefficient of stability vanishes. This, together with the result of the present section, finally establishes that for any given order only the characteristic coefficient of stability can vanish and that it does do so for some finite Jacobi figure.

Again, the argument of p. 93 *et seq.* is available in the present case to show that the equation

$$\frac{1}{3} L_1 S_1 - \frac{1}{2n+1} L_i S_i = 0 \tag{2}$$

has only a single finite root. Now the initial (spheroidal) Jacobi form has $a : b : c = 1{\cdot}197 : 1{\cdot}197 : 0{\cdot}698$ and therefore $\tau^2 = 0{\cdot}515$. The final Jacobi form has $a : b : c = \infty : 0 : 0$ and hence $\tau^2 = 0$. For intermediate values the ratio $a : b$ is uniquely settled if the ratio $a : c$ is assigned (Table II, p. 41) so that equation (2) is simply an equation for $a : c$. But the argument of Chapter VII (p. 95), applies generally to show that the number of finite roots of (2) is *one*. Accordingly, as the Jacobi series is described in the direction of increasing angular momentum, each characteristic coefficient of stability changes sign but once. If displacements of a given order n only are contemplated, the mass eventually becomes secularly unstable for some finite value of $a : b : c$, and thereafter remains secularly unstable.

It can be established (see below) that instability first enters through the characteristic third-order harmonic deformation, and that as the series continues to be described, further instabilities enter successively through the characteristic harmonic of the fourth order, then of the fifth order, next of the sixth order, and so on. If for a given order n we denote the characteristic coefficient of stability by C_n, then the following result can be proved, establishing that these coefficients vanish successively.

If $n > p$, then $C_n > C_p$.

If $C_n = 0$ we have

$$C_n = \frac{1}{3} L_1 S_1 - \frac{1}{2n+1} L_n S_n = 0 \quad (\lambda = 0),$$

where L_n denotes the Lamé function corresponding to the characteristic coefficient of stability of order n. To establish the result it will be sufficient if it can be shown that L_n/L_p, where $n > p$, is always increasing as $\lambda + a^2$ increases from $a^2 - c^2$ to ∞. For if this is so, then

$$\frac{d}{d\lambda}(L_n^2/L_p^2) > 0,$$

and hence (from Chapter VII, p. 93)

$$\frac{d}{d\lambda}\left\{\left(\frac{1}{2p+1} L_p S_p - \frac{1}{2n+1} L_n S_n\right)\Big/ L_p^2\right\} < 0. \tag{5}$$

Or, in terms of C_p and C_n, $\quad \frac{d}{d\lambda}\{(C_n - C_p)/L_p^2\} < 0.$

In this, L_p^2 is essentially positive, and since

$$C_n - C_p \to \left(\frac{1}{2p+1} - \frac{1}{2n+1}\right)\lambda^{-\frac{1}{2}} \to +0 \quad \text{if} \quad n > p \quad \text{as} \quad \lambda \to \infty,$$

this means that $C_n - C_p$ is always positive, and hence that C_p must vanish first.

To show that L_n/L_p is always increasing we consider three cases as follows.

(i) *Suppose $a = b$.*

Since the L-function giving the characteristic coefficient of stability has all its roots in $\lambda + a^2$ between 0 and $a^2 - b^2$, when a becomes equal to b all these roots must come into coincidence at $\lambda + a^2 = 0$, and hence we have

$$n \text{ even} \quad L_n = (a^2 + \lambda)^{\frac{1}{2}n},$$

$$n \text{ odd} \quad L_n = \sqrt{(a^2 + \lambda)} \dots (a^2 + \lambda)^{\frac{1}{2}(n-1)} = (a^2 + \lambda)^{\frac{1}{2}n} \quad \text{again.}$$

Hence $$L_n/L_p = (a^2 + \lambda)^{\frac{1}{2}(n-p)},$$

and since $n > p$, this is always increasing as $a^2 + \lambda$ increases from $a^2 - c^2$ to ∞.

(ii) *Suppose $b = c$.*

In this case it has been seen that $L_n = D^n(t^2 - 1)^n$, and hence

$$\frac{L_n}{L_p} = \frac{D^n(t^2 - 1)^n}{D^p(t^2 - 1)^p},$$

and as λ increases, $t^2 = (a^2 + \lambda)/(a^2 - c^2)$ increases from 1 to ∞. To prove that L_n/L_p increases, if we put $t = 1 + 2x$, then apart from positive factors,

$$L_n/L_p = D^n\{x^n(x+1)^n\}/D^p\{x^p(x+1)^p\}. \tag{6}$$

If $x = 0$ this has the value $n!/p!$, and it is easily shown that

$$p! \, D^n\{x^n(x+1)^n\} - n! \, D^p\{x^p(x+1)^p\}$$

contains only terms with positive coefficients if $n > p$. It is therefore always increasing with $x \; (> 0)$, and hence, so is L_n/L_p.

(iii) *Suppose b arbitrary and between a and c.*

When u (p. 79) is taken as independent variable instead of λ the Lamé equations for L_n and L_p become

$$\frac{d^2 L_n}{du^2} = \{n(n+1)(\lambda + a^2) - K'_n\}L_n, \tag{7}$$

$$\frac{d^2 L_p}{du^2} = \{p(p+1)(\lambda + a^2) - K'_p\}L_p, \tag{8}$$

(the constants K'_n, K'_p have been so written, with dashes, to allow for $\lambda + a^2$ being taken as variable on the right instead of simply λ). Now in order that the ratio L_n/L_p shall *not* be always increasing it is necessary (as has been shown in Chapter VII, p. 94) that the equation in $\lambda + a^2$

$$\psi \equiv \{n(n+1) - p(p+1)\}(\lambda + a^2) - (K'_n - K'_p) = 0$$

shall have a root greater than $a^2 - c^2$, and this can be shown to be impossible by the following argument.

We have (from p. 79),

$$\frac{dL_n}{du} = \frac{dL_n}{d\lambda} \cdot \frac{d\lambda}{du} = -2\sqrt{[(\lambda+a^2)(\lambda+b^2)(\lambda+c^2)]} \frac{dL_n}{d\lambda}. \tag{9}$$

Hence the expression $\qquad L_p \dfrac{dL_n}{du} - L_n \dfrac{dL_p}{du}$

vanishes for $\lambda + a^2 = a^2 - b^2$ and for $\lambda + a^2 = a^2 - c^2$, since $\lambda + b^2 = 0$ and $\lambda + c^2 = 0$ make the radical factor vanish, and L_n itself does not contain $\sqrt{(\lambda+b^2)}$ or $\sqrt{(\lambda+c^2)}$ as a factor. But we have, by means of the differential equations for L_n and L_p,

$$\psi L_n L_p = L_p \frac{d^2 L_n}{du^2} - L_n \frac{d^2 L_p}{du^2}$$

$$= \frac{d}{du}\left(L_p \frac{dL_n}{du} - L_n \frac{dL_p}{du}\right), \tag{10}$$

and hence $\psi L_n L_p$ must vanish at least once in the interval

$$a^2 - b^2 < \lambda + a^2 < a^2 - c^2.$$

But $L_n L_p$ cannot vanish, since all the zeros of both L_n and L_p are less than $a^2 - b^2$. Hence ψ itself must vanish for some $\lambda + a^2$ in the interval, and since ψ is linear it cannot also vanish elsewhere.

Accordingly it is established that in all cases L_n/L_p is always increasing, so that $C_n > C_p$ if $n > p$, and it follows that the coefficients of stability corresponding to harmonics of order 3, 4, 5, ... vanish successively as the Jacobi series is described in the direction of increasing angular momentum.

The L-function corresponding to the characteristic coefficient

To find the characteristic coefficient of any given order n, use can be made of the property that when $b = c$ the corresponding L-function reduces to $D^n(t^2-1)^n$.

This result can be established independently from the fact that in the general case L must be such that all its roots are less than $a^2 - b^2$. For if we put $b = c$, roots $< a^2 - b^2$ will remain so. But then L reduces to

$$L = (t^2-1)^{\frac{1}{2}p} D^{p+n}(t^2-1)^n, \tag{11}$$

and unless $p = 0$ there would be a root at $t^2 = 1$, i.e. $\lambda + a^2 = a^2 - c^2 > a^2 - b^2$ ($p = 1$ is excluded since it would give $\sqrt{(\lambda+c^2)}$ as factor).

For example, for $n = 3$, we require to find the general third-order Lamé function that reduces when $b = c$ to

$$D^3(t^2-1)^3 = t(5t^2-3) = \sqrt{(a^2+\lambda)(\lambda + \tfrac{2}{5}a^2 + \tfrac{3}{5}c^2)},$$

apart from constant factors. The third-order Lamé functions were found on page 60, and it is easily verified that the appropriate one of these so reducing is

$$L_3 = \sqrt{(\lambda+a^2)}\{\lambda + \tfrac{1}{5}(a^2+2b^2+2c^2) + \tfrac{1}{5}(a^4+4b^4+4c^4-7b^2c^2-c^2a^2-a^2b^2)^{\frac{1}{2}}\}.$$

The Jacobi ellipsoid at which the third-order characteristic coefficient of stability vanishes has been determined by Darwin,[*] who finds

$$a : b : c = 1\cdot 8858 : 0\cdot 8150 : 0\cdot 6507$$

and $\qquad \omega^2/2\pi G\rho = 0\cdot 1420, \quad H = 0\cdot 3896.$ (See Table II, p. 41.)

We may conclude therefore that all ellipsoidal forms less elongated than this are thoroughly stable, that is, both secularly and ordinarily stable, and all ellipisoids more elongated are secularly unstable at least for third-order harmonic displacments.

The pear-shaped figure

According to the theory of linear series of equilibrium configurations, the occurrence of a point of bifurcation on the ellipsoidal series indicates that another linear series of equilibrium configurations crosses it at this point. So long as the free surface of the liquid mass is restricted to ellipsoidal form, that is, so long as ξ/ϖ depends only on second-order harmonics, no point of bifurcation can reveal itself, just as the Maclaurin series is found to be always secularly stable if spheroidal forms only are permitted. Moreover, the initial form of the new series can be obtained from the ellipsoidal series by applying a small normal deformation ξ/ϖ depending on the particular surface harmonic $M_3 N_3$ through which instability enters, just as in the spheroidal case the initial members of the Jacobi series can be obtained by means of a small deformation proportional to that $M_2 N_2$ for which the instability enters.

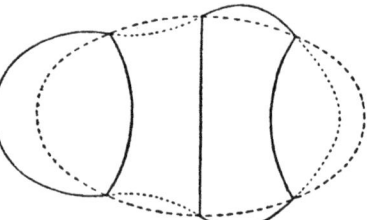

If ξ/ϖ is taken small but finite, indications of the shape of the members of the new series can be calculated. In his original *Mémoir* on this subject, Poincaré gave a sketch of the resulting surface, and although the figure is not of course

Fig. 16. (*After Poincaré, 'Figures d'Equilibre'*, p. 161.)

axially symmetrical round the longest axis (of the original ellipsoid, for which in any case $b \neq c$) the resemblance to a pear was so striking that it seemed appropriate to call the new form 'the pear-shaped figure' and the series 'the pear-shaped series'. Since then it has also been referred to as 'the piriform figure' and sometimes simply 'the pear'. It is of interest, however, that Darwin's detailed calculations, made later, both of the axes of the critical ellipsoid and of the normal displacements corresponding to the crucial third-order harmonic lead to a form of the figure distinctly more elongated than Poincaré had supposed and bearing rather less resemblance to a pear. Poincaré's sketch and the form as found by Darwin are shown in Figs. 16 and 17, which make clear the extent of the resemblance to a pear. But a name is useful in any case for the new series and there has been general acceptance and usage of the term pear-shaped.

[*] *Collected Works*, III, p. 312.

The stability of the pear-shaped figure

The problem of the stability of the pear-shaped figures has been studied by Liapounoff, Darwin, and Jeans, and has proved to be one of the utmost intricacy. The chief difficulty lies in calculating with sufficient accuracy the gravitational potential of the pear-shaped form, the value of which is necessary for the expression of the constancy of the total mechanical potential (gravitational plus centrifugal) over its boundary surface. The pear-shaped figure is not representable

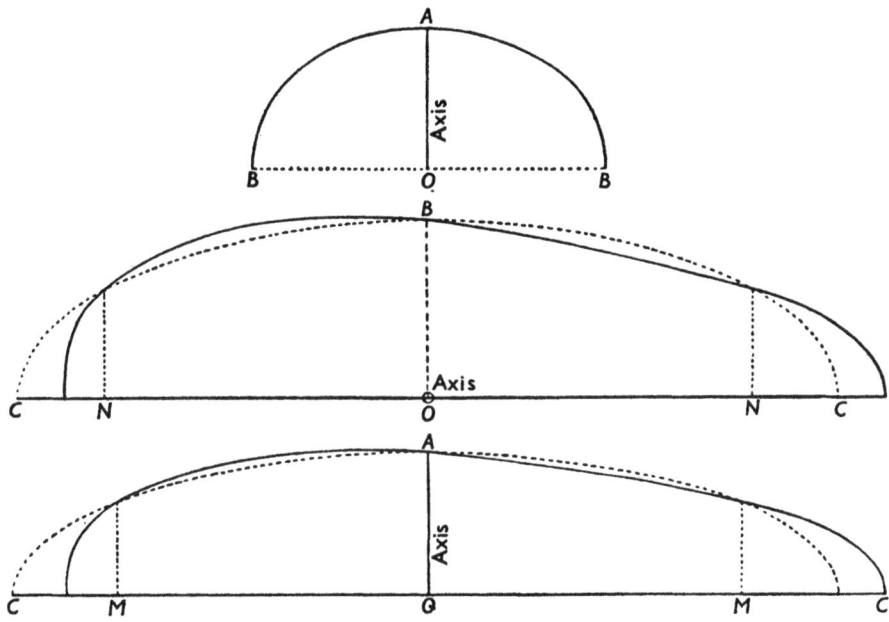

Fig. 17. (*After Darwin, 'Scientific Papers', III, p. 314.*)

by any simple closed analytical form, but it has been found possible to examine its properties in the immediate neighbourhood of the critical Jacobi form at which the piriform series bifurcates, for this can be achieved by regarding the pear-shaped series as produced by the addition to the critical ellipsoid of a small deformation measured by an infinitesimal parameter in powers of which series expansions can be introduced. Thus the equation of the pear-shaped figure can be shown to be capable of being represented to sufficient accuracy by an expression of the form

$$\left(\frac{x^2}{a^2}+\frac{y^2}{b^2}+\frac{z^2}{c^2}-1\right)+eP_0+e^2Q_0+e^3R_0 = 0, \tag{12}$$

where the first term (in brackets) equated to zero is the ellipsoid of bifurcation, e is the small parameter representing the degree of (small) departure from the point of bifurcation, and P_0, Q_0, R_0 are certain polynomials in x, y, z in which the terms of highest degree are respectively of order 3, 4, and 5. The problem is then to calculate the coefficients in these polynomials appropriate to the pear-shaped equilibrium form by means of the condition for relative hydrostatic equilibrium. In the present case this condition is found to take the form

$$V_g + \tfrac{1}{2}\{\omega^2 + e^2\,\delta(\omega^2)\}\,(x^2 + y^2)$$

$$= -\pi G\rho\, abc\, \theta\left(\frac{x^2}{a^2} + \frac{y^2}{b^2} + \frac{z^2}{c^2} - 1 + eP_0 + e^2 Q_0 + e^3 R_0\right). \tag{13}$$

Here V_g is the gravitational potential of the mass bounded by (12), and it turns out to be necessary to calculate V_g as far as terms of the third order in e, that is, to e^3. It is this requirement that renders the problem of such great length and complexity. The term $e^2\,\delta(\omega^2)$ represents the adjustment to the square of the angular velocity, ω^2, associated with the critical Jacobi form, and its value is found in the process of calculating the coefficients in P_0, Q_0, R_0. The factor θ is simply a constant of proportionality included here since it is only necessary that the total potential shall be constant over the free surface.

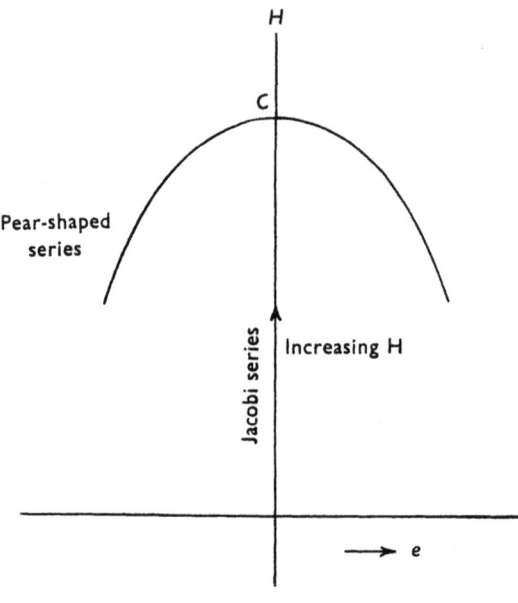

Fig. 18.

When the coefficients involved in P_0, Q_0, R_0 have been found, it is then comparatively easy to calculate the moment of inertia of the corresponding pear-shaped mass as far as terms in e^2, which is all that is finally necessary. Then from the calculated value of the angular velocity $\omega^2 + e^2\,\delta(\omega^2)$, the value of the angular momentum appropriate to the equilibrium pear-shaped forms in the neighbourhood of the point of bifurcation follows at once. According to Jeans's calculations the values are as follows:

$$\omega^2 + e^2\,\delta(\omega^2) = 2\pi G\rho \times 0\cdot142(1 + 0\cdot05227e^2) \tag{14}$$

for the angular velocity of the equilibrium pear-shaped figure, and

$$H = H_0(1 - 0\cdot06765e^2) \tag{15}$$

for its angular momentum, where $H_0 = 0\cdot3896(G^{\frac{1}{2}} M^{\frac{5}{3}} r^{\frac{1}{2}})$, $r^3 = abc$.

If therefore, as usual, we regard the system as evolving slowly in the direction of increasing angular momentum, and represent the Jacobi series by a vertical line in a diagram, and the parameter e, representing departure from it, by a horizontal coordinate, as in Fig. 18, then it is evident from (15) that the pear-shaped series of equilibrium configurations are such that their representative curve initially turns downwards from the point of bifurcation C, as indicated in the diagram. Accordingly, by the considerations of Chapter II (p. 10), the pear-shaped series must be secularly unstable initially. This conclusion has also been reached independently by Liapounoff and, according to Jeans,* can also in fact be established from Darwin's researches once a certain technical error is put right.

What the result means physically is that if the mass were set in motion in precisely the state consistent with a pear-shaped figure (in the neighbourhood of C), it would be in equilibrium, and if entirely undisturbed would continue in rigid-body rotation in that form. Also, its total angular momentum would be slightly less than that associated with the critical Jacobi figure, but its angular velocity slightly faster. If, however, any slight disturbance were given to the system, the presence of internal frictional forces would cause at least some of the amplitudes associated with the disturbance to increase, and the system would gradually depart more and more from the pear-shaped form to settle down finally in that Jacobi form possessing the same angular momentum as the original pear-shaped configuration. It is therefore clear that by no natural course of evolution can any of the initial members of the pear-shaped series ever come into existence. As H gradually increases the mass will evolve along the Jacobi series to C, and if it should increase further the only equilibrium form available for the system will be the appropriate Jacobi ellipsoid, which itself will at this stage have already become secularly unstable.

The nature of the further development of the mass will then depend on whether these ellipsoids are ordinarily stable or not. If they were ordinarily stable, it would mean that any slight disturbance would increase only at a rate dependent on the amount of friction present, and this might not involve a rapid departure. An example of such a case is provided by the lunar orbit which, though secularly unstable, is ordinarily stable, and the rate of departure from the present configuration is extremely slow because of the very small amount of tidal friction available. On the other hand if the Jacobi ellipsoids beyond the point of bifurcation are ordinarily unstable, as in fact they will be shown to be, the exponential factors indicating the instability are independent of friction and do not vanish with it. They accordingly may involve dynamical changes from the Jacobi form proceeding at an acceleration comparable with any other acceleration in the system. The examination of the ordinary stability or otherwise of the Jacobi ellipsoids accordingly forms the subject of the following chapter.

* *Problems of Cosmogony*, p. 92.

Chapter IX

THE ORDINARY STABILITY OF THE JACOBI ELLIPSOIDS

To investigate *ordinary stability* it is necessary to set up the equations of small motion of the liquid and consider the conditions under which their solutions are purely periodic or otherwise.

General equations of small motion of a rotating liquid

We adopt rectangular axes Ox, Oy, Oz and suppose them rotating round Oz with angular velocity ω, which may be assumed constant since we intend to examine small motion about an equilibrium configuration having angular velocity ω. The velocity components u, v, w relative to these axes then suffice to deal with any small departure.

The component accelerations, relative to fixed axes coinciding with the instantaneous position of the moving axes, are

$$\frac{Du}{Dt} - 2\omega v - \omega^2 x, \quad \frac{Dv}{Dt} + 2\omega u - \omega^2 y, \quad \frac{Dw}{Dt},$$

where

$$\frac{D}{Dt} = \frac{\partial}{\partial t} + u\frac{\partial}{\partial x} + v\frac{\partial}{\partial y} + w\frac{\partial}{\partial z}.$$

In small motion all squares and products of u, v, w and their derivatives are neglected, and the component accelerations accordingly simplify to

$$\frac{\partial u}{\partial t} - 2\omega v - \omega^2 x, \quad \frac{\partial v}{\partial t} + 2\omega u - \omega^2 y, \quad \frac{\partial w}{\partial t}.$$

The equations of small motion are therefore

$$\frac{\partial u}{\partial t} - 2\omega v - \omega^2 x = \frac{\partial \phi}{\partial x} - \frac{1}{\rho}\frac{\partial p}{\partial x},$$

$$\frac{\partial v}{\partial t} + 2\omega u - \omega^2 y = \frac{\partial \phi}{\partial y} - \frac{1}{\rho}\frac{\partial p}{\partial y},$$

$$\frac{\partial w}{\partial t} = \frac{\partial \phi}{\partial z} - \frac{1}{\rho}\frac{\partial p}{\partial z},$$

where p denotes the pressure and ϕ is the gravitational potential of the liquid mass, no external forces other than gravitation being supposed present. Also, uniform density is assumed throughout, so writing

$$\chi \equiv p/\rho - \phi - \tfrac{1}{2}\omega^2(x^2 + y^2)$$

these equations take the form

$$\left.\begin{aligned}
\frac{\partial u}{\partial t} - 2\omega v &= -\frac{\partial \chi}{\partial x}, \\[2mm]
\frac{\partial v}{\partial t} + 2\omega u &= -\frac{\partial \chi}{\partial y}, \\[2mm]
\frac{\partial w}{\partial t} &= -\frac{\partial \chi}{\partial z}.
\end{aligned}\right\} \tag{1}$$

In addition we have for uniform density the equation of continuity

$$\frac{\partial u}{\partial x} + \frac{\partial v}{\partial y} + \frac{\partial w}{\partial z} = 0. \tag{2}$$

In any configuration of equilibrium relative to the rotating frame, u, v, w would all be zero, and hence

$$\frac{\partial \chi}{\partial x} = \frac{\partial \chi}{\partial y} = \frac{\partial \chi}{\partial z} = 0.$$

Thus
$$\chi \equiv p/\rho - \phi - \tfrac{1}{2}\omega^2(x^2 + y^2) = \chi_0 \tag{3}$$

where, in the first instance, χ_0 is independent of space coordinates. But χ_0 must also be independent of time, since in a natural equilibrium configuration p and ϕ depend only on the space coordinates x, y, z.

Supposing now that there exists some configuration of equilibrium of the system, it is proposed to consider the possibility of motions in which the maximum displacement of any particular element of the liquid from its mean position is a quantity of the first order of smallness at most, since the object is to arrive at valid linear equations. The resulting non-equilibrium configuration at any instant will be referred to as the disturbed configuration, and will differ in dimensions from the equilibrium configuration only by small amounts of the first order. In the usual manner of treating such motions for the purpose of discussing their stability, it is supposed in building up the relevant equations that each element of the liquid undergoes an oscillation about a mean position, and the reality or otherwise of the period of this motion, when found, deter·mines whether or not the postulated motion is capable of remaining small. Thus, if x_0, y_0, z_0 is the mean position of the element of liquid that at time t has position x, y, z, then we may suppose

$$\left.\begin{aligned}
x - x_0 &= \mathfrak{R}(\xi\, e^{i\lambda t}), \\[1mm]
y - y_0 &= \mathfrak{R}(\eta\, e^{i\lambda t}), \\[1mm]
z - z_0 &= \mathfrak{R}(\zeta\, e^{i\lambda t}),
\end{aligned}\right\} \tag{4}$$

where $\mathfrak{R}$ denotes the real part of the functions following it. ($i = \sqrt{-1}$.) The period $2\pi/\lambda$ connected with the motion is the same for every particle of the liquid in accordance with the usual procedure of finding small oscillations. (There will be no confusion of the present λ with that of the confocal parameters.) The quantities ξ, η, ζ will be functions of position, and as $x - x_0$, etc. are small, so must be the magnitudes of these functions. It is therefore immaterial to the present order of accuracy whether they are regarded as functions of x_0, y_0, z_0 or of x, y, z. As they stand in (4) they will in general involve complex coefficients owing to the factor $e^{i\lambda t}$, and it will be supposed without continuing to write the sign $\Re$ that the real parts give the appropriate displacements.

Whatever form these quantities ξ, η, ζ take, it is to be expected that as in ordinary dynamical examples of small oscillations, they will be undetermined to within one or more (infinitesimal) factors that measure the amplitudes, as it were, of the oscillations. If this factor (or factors) is reduced to zero, no oscillation would take place, since all amplitudes involved are linear multiples of it. Thus (x_0, y_0, z_0) must in fact be the position of the element in the equilibrium configuration, and accordingly in the motion considered every element oscillates harmonically (if λ is real) about its position in the assumed equilibrium state. Thus x_0, y_0, z_0 refer to the equilibrium configuration.

For such a motion equations (4) require to the first order that

$$u \equiv \frac{\partial x}{\partial t} = i\lambda \xi e^{i\lambda t} \quad \text{and} \quad \frac{\partial u}{\partial t} = -\lambda^2 \xi e^{i\lambda t},$$

with similar expressions for v, $\dfrac{\partial v}{\partial t}$; w and $\dfrac{\partial w}{\partial t}$. Inserting these in (1), a small motion of this kind requires

$$\left. \begin{aligned} \frac{\partial \chi}{\partial x} &= (\lambda^2 \xi + 2i\omega\lambda\eta)\, e^{i\lambda t}, \\[2mm] \frac{\partial \chi}{\partial y} &= (\lambda^2 \eta - 2i\omega\lambda\xi)\, e^{i\lambda t}, \\[2mm] \frac{\partial \chi}{\partial z} &= \lambda^2 \zeta\, e^{i\lambda t}. \end{aligned} \right\} \tag{5'}$$

These equations show that if χ contains t in any other form than $e^{i\lambda t}$ it cannot do so in any terms involving x, y, or z. But any terms not containing x, y, z can be omitted, since only space differentiations of χ enter the equations of motion. Thus we may write

$$\chi = \Re\{\psi(x, y, z)\, e^{i\lambda t}\} + (\chi_0) \tag{4'}$$

where, again, ψ may have complex coefficients, and the real part of the right-hand side gives the χ appropriate to the motion. Thus if the oscillation is reduced to zero, χ must become χ_0, and hence ψ must reduce to zero.

Equations (5'), together with the equation of continuity, now take the form

$$\left. \begin{aligned} \frac{\partial \psi}{\partial x} &= \lambda^2 \xi + 2i\omega\lambda\eta, \\[2mm] \frac{\partial \psi}{\partial y} &= \lambda^2 \eta - 2i\omega\lambda\xi, \\[2mm] \frac{\partial \psi}{\partial z} &= \lambda^2 \zeta, \end{aligned} \right\} \tag{5}$$

and

$$\frac{\partial \xi}{\partial x} + \frac{\partial \eta}{\partial y} + \frac{\partial \zeta}{\partial z} = 0.$$

The small oscillations of a general rotating liquid are thus seen to depend on these four partial differential equations. The functional forms of ξ, η, ζ, and ψ, as is usual in such circumstances, will depend on the boundary conditions obtaining in the particular problem to which the equations are to be applied.

As already mentioned, the expressions giving the motion must, as for systems of a finite number of degrees of freedom, contain one or more (independent) arbitrary infinitesimal constants determining the general amplitude of the motion, in terms of which all other constants concerned, for each particular λ, must be capable of being linearly expressed, and if this one constant is made to vanish so must the others. The equations are in fact seen to be satisfied by $\xi = \eta = \zeta = \psi = 0$, and if ξ, η, ζ, ψ is a solution, so also is $k\xi, k\eta, k\zeta, k\psi$ for any small constant k. The equation determining the periods, or λ, is to be obtained, as for finite dynamical systems, by eliminating all the constants associated with the various amplitudes of the oscillation.

The surface condition for an ellipsoidal configuration

If the liquid is bounded by a free surface, as in the problem to be considered, then $p = 0$ at this surface whatever its form. From (3) the surface value of χ is therefore

$$\chi_s = -\phi - \tfrac{1}{2}\omega^2(x^2 + y^2),$$

where now ϕ is the gravitational potential of the disturbed configuration. To calculate ϕ at a general point, the disturbed configuration may be regarded as consisting of the equilibrium configuration together with a layer of liquid spread on its surface and of depth PN (Fig. 19) given by

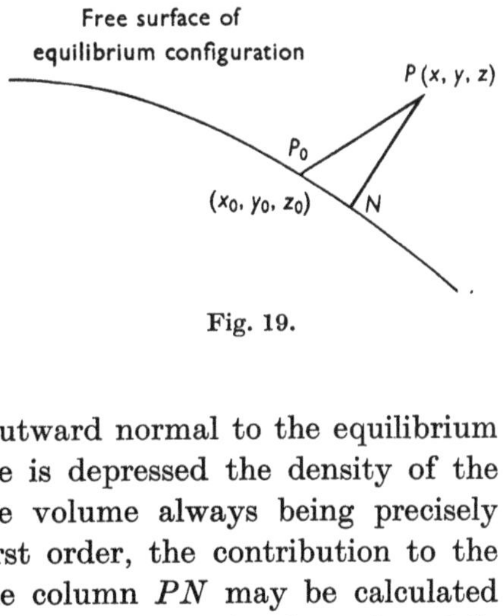

$$PN = l(x - x_0) + m(y - y_0) + n(z - z_0),$$

Fig. 19.

where l, m, n are the direction cosines of the outward normal to the equilibrium surface. (At places where the original surface is depressed the density of the layer must be considered negative, its whole volume always being precisely zero.) Since PP_0 is an infinitesimal of the first order, the contribution to the potential (at a general point of space) by the column PN may be calculated as if it were all situated at P_0 with a resulting error of second order only, while the self-gravitational energy of the layer is also of the second order of smallness. The surface layer may therefore be dealt with, to the first order, by regarding a mass of amount

$$\rho(l\xi + m\eta + n\zeta)e^{i\lambda t} \quad \text{per unit area}$$

as added at the point (x_0, y_0, z_0). Let us denote the potential due to this layer by $\phi_L . e^{i\lambda t}$.

While the foregoing simplification is possible in calculating the potential of an infinitesimal column surrounding PN, it is seen that a similar simplification cannot be made in formulating the condition expressing the vanishing of the pressure at the free surface. For when an oscillation is present, this free surface is no longer at (x_0, y_0, z_0) but at some point (x, y, z). Hence the contribution to the total mechanical potential by the gravitation of the undisturbed configuration plus the centrifugal force must be evaluated at (x, y, z) in order to obtain the value of p correct to first-order accuracy. The gravitational potential of the surface layer can, on the other hand, be evaluated at (x_0, y_0, z_0).

If $-\chi_E$ is the total mechanical potential at a point (x_0, y_0, z_0) of the free surface in the equilibrium configuration, the potential at (x, y, z) in the disturbed configuration will be

$$-\chi_E - \frac{\partial \chi_E}{\partial x_0}(x - x_0) - \frac{\partial \chi_E}{\partial y_0}(y - y_0) - \frac{\partial \chi_E}{\partial z_0}(z - z_0) + \phi_L \cdot e^{i\lambda t},$$

so that

$$-\chi(x, y, z, t) = -\chi_E - \left(\xi \frac{\partial \chi_E}{\partial x_0} + \eta \frac{\partial \chi_E}{\partial y_0} + \zeta \frac{\partial \chi_E}{\partial z_0} \right) e^{i\lambda t} + \phi_L e^{i\lambda t}.$$

But, as explained above, χ_E is constant over the equilibrium free surface (and indeed throughout its interior) and equal to χ_0. Hence, since $\chi = \psi e^{i\lambda t} + \chi_0$, we have

$$\psi(x, y, z, t) = \xi \frac{\partial \chi_E}{\partial x_0} + \eta \frac{\partial \chi_E}{\partial y_0} + \zeta \frac{\partial \chi_E}{\partial z_0} - \phi_L.$$

Now $\dfrac{\partial \chi_E}{\partial x_0}, \dfrac{\partial \chi_E}{\partial y_0}, \dfrac{\partial \chi_E}{\partial z_0}$ are simply the components of total force (gravitational plus centrifugal) at the surface of the equilibrium configuration and are therefore the components of (apparent) surface gravity. Thus if $g(x_0, y_0, z_0)$ denotes the value of gravity

$$\frac{\partial \chi_E}{\partial x_0} : \frac{\partial \chi_E}{\partial y_0} : \frac{\partial \chi_E}{\partial z_0} : g(x_0, y_0, z_0) = l : m : n : 1$$

and hence

$$\psi(x, y, z) = g(l\xi + m\eta + n\zeta) - \phi_L. \tag{6}$$

Assuming now that the undisturbed free surface is the surface of the ellipsoid

$$\frac{x^2}{a^2} + \frac{y^2}{b^2} + \frac{z^2}{c^2} = 1 \quad (a^2 > b^2 > c^2)$$

then

$$l : m : n : 1 = \frac{px_0}{a^2} : \frac{py_0}{b^2} : \frac{pz_0}{c^2} : 1,$$

where

$$\frac{1}{p^2} = \frac{x_0^2}{a^4} + \frac{y_0^2}{b^4} + \frac{z_0^2}{c^4}.$$

It will be recalled that $p = \varpi abc$, where ϖ is the quantity defined on page 77.

Writing now

$$\sigma = \frac{x_0}{a^2}\xi + \frac{y_0}{b^2}\eta + \frac{z_0}{c^2}\zeta = \frac{(l\xi + m\eta + n\zeta)}{p}, \tag{7}$$

then we have, rewriting (6), $\psi = gp\sigma - \phi_L. \tag{8}$

(From this point onwards the suffix can be dropped from x_0, y_0, z_0.)

But it was shown (p. 85) that the product $g\varpi$, and hence gp, is constant over the surface of the ellipsoid and given by

$$gp = \tfrac{4}{3}\pi G\rho\, abc\, L_1(\lambda=0)\, S_1(\lambda=0),$$

wherein $\qquad L_1 = \sqrt{(\lambda+c^2)},$

$$S_1 = \sqrt{(\lambda+c^2)}\int_\lambda^\infty \frac{3d\lambda}{2(\lambda+c^2)\sqrt{[(a^2+\lambda)(b^2+\lambda)(c^2+\lambda)]}},$$

and $\lambda = 0$ is of course the ellipsoid under consideration. The factor $G\rho$ may conveniently be taken as unity.

To arrive at the surface condition in an appropriate form it is only necessary now to express σ in terms of surface harmonics. The quantities ξ, η, ζ are functions of position on the surface of the ellipsoid, and hence so is σ. It can therefore in general be developed in a series of Lamé functions (involving of course only μ, ν) thus

$$\sigma = \sum_k A_k M_k N_k \tag{9}$$

where the A_k's are (small) constants. The potential ϕ_L due to a surface layer of depth $p\sigma$ spread on the ellipsoid may now be written down at once from page 82, and is

$$\phi_L = 4\pi abc\, \Sigma A_k \cdot \frac{L_k(0)\, S_k(0)}{2n+1}\, M_k N_k, \tag{10}$$

where n is the order of the Lamé functions L_k, M_k, N_k. Hence on the surface of the ellipsoid we have from (8)

$$\psi = 4\pi abc\, \Sigma A_k \left\{ \frac{1}{3}L_1 S_1 - \frac{1}{2n+1}L_k S_k \right\}_{\lambda=0} M_k N_k. \tag{11}$$

For brevity we write

$$H_o = \frac{2}{3}abc\, L_1(0)\, S_1(0) \quad \text{and} \quad H_k = \frac{2abc}{2n+1}\, L_k(0)\, S_k(0).$$

Then the boundary condition to be satisfied by the functions ξ, η, ζ, ψ is seen to be equivalent to the following condition:

If the constants of the ellipsoidal harmonic expansion of the function

$$\sigma = (l\xi + m\eta + n\zeta)/p$$

are denoted by A_k, then the expansion constants of ψ on the ellipsoid must be $2\pi(H_o - H_k)A_k$.

Expression of the boundary condition in terms of ψ and its derivatives

Solving the first three of equations (5) for ξ, η, ζ gives, provided $\lambda(\lambda^2 - 4\omega^2) \neq 0$,

$$\left.\begin{aligned}
\xi &= \left(\frac{\partial\psi}{\partial x} - \frac{2i\omega}{\lambda}\frac{\partial\psi}{\partial y}\right)\Big/(\lambda^2 - 4\omega^2), \\[2mm]
\eta &= \left(\frac{\partial\psi}{\partial y} + \frac{2i\omega}{\lambda}\frac{\partial\psi}{\partial x}\right)\Big/(\lambda^2 - 4\omega^2), \\[2mm]
\zeta &= \frac{\partial\psi}{\partial z}\Big/\lambda^2,
\end{aligned}\right\} \tag{12}$$

and inserting these in the equation of continuity shows that ψ, in space, must satisfy the second-order partial differential equation

$$\nabla_1^2 \psi \equiv \frac{\partial^2 \psi}{\partial x^2} + \frac{\partial^2 \psi}{\partial y^2} + \left(1 - \frac{4\omega^2}{\lambda^2}\right)\frac{\partial^2 \psi}{\partial z^2} = 0. \tag{13}$$

This equation is usually referred to as Poincaré's equation.

We have also, from the definition of σ, that

$$(\lambda^2 - 4\omega^2)\sigma = \frac{x}{a^2}\frac{\partial \psi}{\partial x} + \frac{y}{b^2}\frac{\partial \psi}{\partial y} + \left(1 - \frac{4\omega^2}{\lambda^2}\right)\frac{z}{c^2}\frac{\partial \psi}{\partial z} - \frac{2i\omega}{\lambda}\left(\frac{x}{a^2}\frac{\partial \psi}{\partial y} - \frac{y}{b^2}\frac{\partial \psi}{\partial x}\right)$$

$$= D_\lambda(\psi) \quad \text{say.} \tag{14}$$

It may be remarked here, from the above definition of $D_\lambda(\psi)$, that if ψ is a polynomial of degree n so is $D_\lambda(\psi)$. The problem therefore reduces to the following requirement:

To find a solution ψ, of Poincaré's equation, defined everywhere in and on the ellipsoid and such that on the surface of the ellipsoid ψ and $D_\lambda(\psi)$ have the constants of their ellipsoidal harmonic expansions respectively of the forms

$$2\pi(H_o - H_k)A_k \quad \text{and} \quad (\lambda^2 - 4\omega^2)A_k.$$

Properties of Poincaré's equation

Before proceeding to this problem it is necessary to establish certain properties of Poincaré's equation. These are stated under Theorems I–IV below.

We shall suppose that λ can take any values, real, complex, or pure imaginary, but with the exception of real values in the range $-2\omega \leqslant \lambda \leqslant 2\omega$. That is, the only real values of λ permitted are such that $\lambda^2 > 4\omega^2$ or $1 - 4\omega^2/\lambda^2 > 0$. Arguments based on this restriction can be applied directly to discussing stability, since this may be dealt with by showing that imaginary values of λ are impossible. That is, we can assume imaginary values for λ and show that this leads to a contradiction.

Theorem I. Poincaré's equation possesses no solution that vanishes on the surface of the ellipsoid, other than the identical one $\psi = 0$.

$\left(\text{This result would obviously not be generally true for } 1 - \frac{4\omega^2}{\lambda^2} < 0, \text{ for then}\right.$

$\psi = \dfrac{x^2}{a^2} + \dfrac{y^2}{b^2} + \dfrac{z^2}{c^2} - 1$ would be a solution vanishing on the ellipsoid if λ were such

that $1 - \dfrac{4\omega^2}{\lambda^2} = -\dfrac{1/a^2 + 1/b^2}{1/c^2}.\Bigg)$

To prove this we make use of Green's Lemma, or the so-called divergence theorem,

$$\iint(lU + mV + nW)\,dS = \iiint\left(\frac{\partial U}{\partial x} + \frac{\partial V}{\partial y} + \frac{\partial W}{\partial z}\right)dx\,dy\,dz,$$

in which the integral on the left will be applied to the surface of the ellipsoid and the volume integral to its interior. Denoting by $\bar{\psi}$ the conjugate complex function to ψ a solution of Poincaré's equation, let us take for U, V, W the expressions

$$U = \bar{\psi}\frac{\partial \psi}{\partial x}, \quad V = \bar{\psi}\frac{\partial \psi}{\partial y}, \quad W = \bar{\psi}\Big(1 - \frac{4\omega^2}{\lambda^2}\Big)\frac{\partial \psi}{\partial z}.$$

Inserting these in Green's formula gives

$$\iint \bar{\psi}\Big\{l\frac{\partial \psi}{\partial x} + m\frac{\partial \psi}{\partial y} + n\Big(1 - \frac{4\omega^2}{\lambda^2}\Big)\frac{\partial \psi}{\partial z}\Big\}\,dS = \iiint \Big\{\frac{\partial \psi}{\partial x}\cdot\frac{\partial \bar{\psi}}{\partial x} + \frac{\partial \psi}{\partial y}\cdot\frac{\partial \bar{\psi}}{\partial y} + \Big(1 - \frac{4\omega^2}{\lambda^2}\Big)\frac{\partial \psi}{\partial z}\cdot\frac{\partial \bar{\psi}}{\partial z}\Big\}\,dx\,dy\,dz,$$

the remaining terms, multiplying $\bar{\psi}$ and arising on the right, disappearing since ψ satisfies Poincaré's equation.

If now it is assumed that ψ vanishes on the surface, then so also must $\bar{\psi}$, and hence the left-hand side of the above is zero. Thus in that event the integral on the right must also vanish. But $\frac{\partial \psi}{\partial x}\cdot\frac{\partial \bar{\psi}}{\partial x}$ is real and positive, as also is $\frac{\partial \psi}{\partial y}\cdot\frac{\partial \bar{\psi}}{\partial y}$. Hence if λ^2 is complex, or pure imaginary, the vanishing of the imaginary part of the integral on the right gives

$$\iiint \frac{\partial \psi}{\partial z}\cdot\frac{\partial \bar{\psi}}{\partial z}\,dx\,dy\,dz = 0.$$

But $\frac{\partial \psi}{\partial z}\cdot\frac{\partial \bar{\psi}}{\partial z} \geqslant 0$ essentially everywhere, and therefore, in fact, $\frac{\partial \psi}{\partial z} = 0$. Hence we must also have $\frac{\partial \psi}{\partial x} = \frac{\partial \psi}{\partial y} = 0$.

In the same way, if λ^2 is real and $1 - 4\omega^2/\lambda^2 > 0$, the same result holds, since all three terms on the right are now essentially non-negative. Hence ψ is constant within the ellipsoid, and since it vanishes on the surface it must be zero everywhere.

It follows immediately that Poincaré's equation can have only one solution taking given values on the surface of the ellipsoid. For if ψ_1 and ψ_2 were two essentially different solutions taking the same given values at the surface, then $\psi_1 - \psi_2$ would be a solution taking zero value on the surface. Accordingly $\psi_1 - \psi_2$ would be zero everywhere within the ellipsoid, that is $\psi_1 \equiv \psi_2$.

Theorem II. There always exists a polynomial $P(x, y, z)$ of degree n that satisfies Poincaré's equation and takes on the ellipsoid the same values as an arbitrarily given polynomial $Q(x, y, z)$ of the same degree.

This will certainly be true if it can be shown that there exists a polynomial $R(x, y, z)$ of degree $n - 2$ such that the polynomial defined by

$$P \equiv Q\,(\text{given}) + \Big(\frac{x^2}{a^2} + \frac{y^2}{b^2} + \frac{z^2}{c^2} - 1\Big)R$$

is a solution of the equation, i.e. $\nabla_1^2 P = 0$. (The polynomials involved here are not being supposed homogeneous.) Now R, being of degree $n - 2$, contains $\frac{1}{6}n(n^2 - 1)$ coefficients which are as yet at our disposal. Writing down the condition that $\nabla_1^2 P = 0$ gives

$$\nabla_1^2 Q + \Big(\frac{x^2}{a^2} + \frac{y^2}{b^2} + \frac{z^2}{c^2} - 1\Big)\nabla_1^2 R + 4\sum_{x,\,y,\,z}(\text{const.})\,x\frac{\partial R}{\partial x} + (\text{const.})\,R = 0.$$

The expression on the left-hand side of this equation also has $\frac{1}{6}n(n^2-1)$ coefficients, since it also is of degree $n-2$. Also all the coefficients of R occur only linearly, and hence the conditions for it to vanish identically provide just enough (linear) equations to enable the $\frac{1}{6}n(n^2-1)$ coefficients of R to be uniquely found in general.

Now in order that the process of solution shall be possible, it is necessary that the determinant of the equations for the unknown coefficients in R (that is, of the left-hand sides of the resulting equations for the coefficients of R) shall not be zero. It can easily be shown that this requirement is fulfilled. For if we make the equations homogeneous by putting $Q = 0$ temporarily, the resulting determinant cannot vanish, for if it did, it would mean that each of the coefficients in R could be expressed as a multiple of a selected one of them. That is $\frac{1}{6}n(n^2-1)-1$ of them could be expressed as multiples of the remaining one, simply by omitting one of the equations and solving the remainder. Denoting the selected coefficient by a_r this would then give

$$R = a_r \text{ (polynomial of degree } n\text{–}2 \text{ not identically zero).}$$

But then $\left(\dfrac{x^2}{a^2}+\dfrac{y^2}{b^2}+\dfrac{z^2}{c^2}-1\right)R$ would be a polynomial satisfying $\nabla_1^2\psi = 0$ and vanishing on the ellipsoid. Since by Theorem I this is not possible, it follows that the determinant of the coefficients cannot vanish. Hence given Q, an R can always be found, and so P can be found.

Definition. If Q is chosen to be a Lamé polynomial of order n, viz. of the form

$$\left\{\begin{array}{ccc} & x & yz \\ 1 & y & zx \quad xyz \\ & z & xy \end{array}\right\} \Pi\left(\frac{x^2}{a^2+\delta}+\frac{y^2}{b^2+\delta}+\frac{z^2}{c^2+\delta}-1\right)$$

where $\nabla^2 Q = 0$, the corresponding polynomial P will be called a Poincaré polynomial.

Since, in space, Q will be of the form $L_k M_k N_k$, it follows that both polynomials will reduce on the surface of the ellipsoid to a Lamé surface function $M_k N_k$.

Theorem III. If P is a Poincaré polynomial of degree n, the function $D_\lambda(P)$ reduces on the ellipsoid to a sum of Lamé surface functions $M_k N_k$ of order n only.

To prove this, let it be supposed that ϕ and ψ are any two different solutions of Poincaré's equation. Then we have, p denoting as usual the perpendicular from the centre to the tangent plane at an arbitrary point of the ellipsoid,

$$pD_\lambda(\psi) = \frac{px}{a^2}\frac{\partial\psi}{\partial x}+\frac{py}{b^2}\frac{\partial\psi}{\partial y}+\left(1-\frac{4\omega^2}{\lambda^2}\right)\frac{pz}{c^2}\frac{\partial\psi}{\partial z}-\frac{2i\omega}{\lambda}\left(\frac{px}{a^2}\frac{\partial\psi}{\partial y}-\frac{py}{b^2}\frac{\partial\psi}{\partial x}\right)$$

$$= l\frac{\partial\psi}{\partial x}+m\frac{\partial\psi}{\partial y}+n\left(1-\frac{4\omega^2}{\lambda^2}\right)\frac{\partial\psi}{\partial z}-\frac{2i\omega}{\lambda}\left(l\frac{\partial\psi}{\partial y}-m\frac{\partial\psi}{\partial x}\right),$$

and similarly, changing λ to $-\lambda$,

$$pD_{-\lambda}(\phi) = l\frac{\partial\phi}{\partial x}+m\frac{\partial\phi}{\partial y}+n\left(1-\frac{4\omega^2}{\lambda^2}\right)\frac{\partial\phi}{\partial z}+\frac{2i\omega}{\lambda}\left(l\frac{\partial\phi}{\partial y}-m\frac{\partial\phi}{\partial x}\right),$$

where l, m, n are the direction cosines of the outward normal to the ellipsoid. We then have that the surface integral over the ellipsoid

$$\iint \{\phi D_\lambda(\psi) - \psi D_{-\lambda}(\phi)\}\, p\, dS$$

$$= \iint \left\{ l\left(\phi \frac{\partial \psi}{\partial x} - \psi \frac{\partial \phi}{\partial x} - \frac{2i\omega}{\lambda}\phi \frac{\partial \psi}{\partial y} - \frac{2i\omega}{\lambda}\psi \frac{\partial \phi}{\partial y}\right) \right.$$

$$\left. + m\left(\phi \frac{\partial \psi}{\partial y} - \psi \frac{\partial \phi}{\partial y} + \frac{2i\omega}{\lambda}\phi \frac{\partial \psi}{\partial x} + \frac{2i\omega}{\lambda}\psi \frac{\partial \phi}{\partial x}\right) + n\left(1 - \frac{4\omega^2}{\lambda^2}\right)\left(\phi \frac{\partial \psi}{\partial z} - \psi \frac{\partial \phi}{\partial z}\right) \right\} dS$$

$$= \iiint \left\{ \frac{\partial}{\partial x}\left(\phi \frac{\partial \psi}{\partial x} - \psi \frac{\partial \phi}{\partial x} - \frac{2i\omega}{\lambda}\phi \frac{\partial \psi}{\partial y} - \frac{2i\omega}{\lambda}\psi \frac{\partial \phi}{\partial y}\right) \right.$$

$$\left. + \frac{\partial}{\partial y}\left(\phi \frac{\partial \psi}{\partial y} - \psi \frac{\partial \phi}{\partial y} + \frac{2i\omega}{\lambda}\phi \frac{\partial \psi}{\partial x} + \frac{2i\omega}{\lambda}\psi \frac{\partial \phi}{\partial x}\right) + \left(1 - \frac{4\omega^2}{\lambda^2}\right)\frac{\partial}{\partial z}\left(\phi \frac{\partial \psi}{\partial z} - \psi \frac{\partial \phi}{\partial z}\right) \right\} dx\, dy\, dz$$

$$= \iiint \{\phi \nabla_1^2 \psi - \psi \nabla_1^2 \phi\}\, dx\, dy\, dz$$

$$= 0 \quad \text{since } \phi \text{ and } \psi \text{ satisfy Poincaré's equation.}$$

Let it be supposed now that P is a Poincaré polynomial of degree n, and reducing therefore on the ellipsoid to a product of Lamé functions MN of order n, and let us suppose $\psi = P$. Also let us suppose $\phi = P'$ where P' is any other Poincaré polynomial of degree $m < n$. (The m, n here are of course integers and not the conventional symbols for direction cosines in the surface integrals.) Then by the preceding result

$$\iint P' D_\lambda(P)\, p\, dS = \iint P D_{-\lambda}(P')\, p\, dS.$$

But the right-hand side of this vanishes, for $D_{-\lambda}(P')$ is simply a polynomial of degree $m < n$, and therefore expressible on the ellipsoid as a sum of Lamé functions MN of orders m and less, and therefore of orders all less than n. That is

$$D_{-\lambda}(P') = \Sigma A_k M_k N_k \quad \text{the order of every term being less than } n.$$

Also $\qquad\qquad P = MN \quad$ wherein the order is n.

But by the orthogonality property

$$\iint M N M_k N_k\, p\, dS = 0.$$

Hence $\qquad\qquad\qquad \iint P' D_\lambda(P)\, p\, dS = 0.$

That is, $D_\lambda(P)$ on the ellipsoid is orthogonal to all Lamé functions of order less than n. But in space $D_\lambda(P)$ is a polynomial of degree n, and hence because of this result it must reduce on the ellipsoid to a sum of Lamé functions of order n only. (If any other order were involved the integral could not vanish for every P' of order less than n.)

If therefore the $2n+1$ independent Lamé polynomials LMN of order n are considered, and P_1, $P_2,..., P_{2n+1}$ denote the corresponding Poincaré polynomials, then *on the ellipsoid* there must exist relations of the form

$$D_\lambda(P_i) = h_{i,1}P_1 + h_{i,2}P_2 + ... + h_{i,2n+1}P_{2n+1}$$

$$= \sum_1^{2n+1} h_{i,k}P_k,$$

wherein the constants $h_{i,k}$ depend *rationally* on λ since $D_\lambda(P)$ contains λ only rationally.

Similarly $D_{-\lambda}(P)$ reduces on the ellipsoid to a sum of Lamé functions of degree n and the coefficients are again rational in λ.

Theorem IV. If ψ is any function satisfying Poincaré's equation, the expansion constants of order n of $D_\lambda(\psi)$ considered as a function of position on the surface of the ellipsoid, are linear combinations with fixed coefficients of the expansion constants of order n only of ψ.

To establish this we make use of the result arrived at in proving Theorem III, namely,

$$\iint \phi D_\lambda(\psi)\, p\, dS = \iint \psi D_{-\lambda}(\phi)\, p\, dS.$$

In this let us write $\phi = P$, a Poincaré polynomial of order n, and therefore reducing to a Lamé product MN of order n on the ellipsoid. In space P will correspond to a certain Lamé polynomial LMN also of order n. The corresponding expansion constant in $D_\lambda(\psi)$ will be

$$\iint P D_\lambda(\psi)\, p\, dS,$$

and this is equal to

$$\iint \psi D_{-\lambda}(P)\, p\, dS.$$

But, by Theorem III, $D_{-\lambda}(P)$ reduces on the ellipsoid to a sum of Lamé functions of order n only. Hence this integral is equal to a linear sum of the expansion constants of ψ of order n only, which establishes the result.

From this it follows that if in the development of ψ, considered on the surface, into a sum of Lamé products F_i, say, $= MN$, and we consider the terms of order n, thus

$$\psi = ... + (c_1 F_1 + c_2 F_2 + ... + c_{2n+1} F_{2n+1}) + ...,$$

then to each F_i will correspond a Poincaré polynomial P, and we can consider ψ in space to be given by

$$\psi = ... + (c_1 P_1 + c_2 P_2 + ... + c_{2n+1} P_{2n+1}) + ...$$

Then by Theorem IV

$$D_\lambda(\psi) = \sum_{i,\, k=1}^{2n+1} c_i h_{i,k} P_k \quad \text{in space}$$

$$= \sum_{i,\, k=1}^{2n+1} c_i h_{i,k} F_k \quad \text{on the ellipsoid.}$$

Hence the complete value of $D_\lambda(\psi)$ on the surface will be

$$D_\lambda(\psi) = \ldots + \left(\sum_i c_i h_{i,1} F_1 + \sum_i c_i h_{i,2} F_2 + \ldots \right) + \ldots$$

Now the problem to be solved is to find a solution ψ of Poincaré's equation such that, on the ellipsoid, ψ and $D_\lambda(\psi)$ have the constants of their ellipsoidal harmonic expansions respectively of the forms $2\pi(H_o - H_k)A_k$ and $(\lambda^2 - 4\omega^2) A_k$. In order to satisfy this requirement we thus obtain the following conditions on λ

$$(\lambda^2 - 4\omega^2) A_k = 2\pi \sum_{i=1}^{2n+1} h_{i,k}(H_o - H_i) A_i \qquad (k = 1, 2, \ldots, 2n+1), \tag{15}$$

and these conditions evidently constitute $2n + 1$ homogeneous linear relations between the A_k in which the coefficients $h_{i,k}$ depend rationally on λ.

The expansion constants of σ, i.e. in effect of $D_\lambda(\psi)$, are therefore given by an infinity of such linear and homogeneous equations. But this system breaks up into an infinity of partial systems each consisting of a finite number, $2n + 1$, in this same number of unknowns. That is, the constants corresponding to a given order n are involved in $2n + 1$ homogeneous equations that contain none of the constants corresponding to any other order. The equations for the permissible values of λ are obtained by equating to zero the determinants of each of these partial systems, and accordingly lead for each separate ·order n to a rational algebraic equation for λ of a finite degree.

It is shown later that the degree of the equation in λ corresponding to order n is always $n^2 + 4n + 1$. The roots of this equation then give the frequencies, $\lambda/2\pi$, of the possible (small) free oscillations, and then for any selected root, λ, equations (15) give the corresponding ratios of coefficients $A_1 : A_2 : \ldots : A_{2n+1}$ associated with the harmonics in the oscillation. It has been further shown by Cartan that the procedure explained above in fact applies for all frequencies including also any within the range -2ω to 2ω.

Ordinary stability

It will not be necessary to our purpose to obtain in detail expressions for the small oscillations of the Jacobi ellipsoids when these are secularly stable and therefore also ordinarily stable, but to consider only the question of whether they remain ordinarily stable beyond the configuration at which secular stability first ceases. It is assumed throughout, as its definition, that ordinary stability means that all the roots in λ must be real, for if there were an imaginary or complex root there would be a motion in which the displacement (measured to the first order) increases indefinitely and therefore, at least, could not be regarded as remaining small. Moreover, it is not possible in the present case for a motion depending on a term such as e^{-kt} to occur without there also being a term e^{kt}. For to each solution of equations (5) there exists a corresponding solution symmetrical with respect to the xz-plane but with the sign of λ changed, since the equations are unaltered if the signs of y, η and λ are changed. This shows that the roots of the final equation in λ derived from (5) are equal and opposite in pairs, as might have been expected but cannot be asserted from the general result for finite rotating systems.

If the ellipsoid is secularly stable for displacements of order n, the potential energy will be an absolute minimum and therefore the system will automatically be ordinarily stable, and so the equation in λ will necessarily have all its roots real. Accordingly, as the system evolves along the Jacobian series in the direction of increasing angular momentum, the question of ordinary instability first arises for third-order harmonic deformations. *It will next be shown that ordinary stability ceases at the same time as the secular stability for these displacements.* From a physical standpoint, this property amounts to the Jacobian ellipsoids being ordinarily unstable at all stages beyond the critical form at which the pear-shaped series bifurcates.

To establish this it is necessary to consider in further detail the actual equation for the possible values of λ, but it turns out that the result depends only the highest and lowest terms of the equation, and we proceed to consider these.

The degree of the equation for λ for oscillations of order n

It has been seen that the small oscillations are to be obtained by integrating the equations

$$\left.\begin{aligned}\frac{\partial \psi}{\partial x} &= \lambda^2 \xi + 2i\omega\lambda\eta, \\[1mm] \frac{\partial \psi}{\partial y} &= \lambda^2 \eta - 2i\omega\lambda\xi, \\[1mm] \frac{\partial \psi}{\partial z} &= \lambda^2 \zeta. \end{aligned}\right\} (a) \qquad \frac{\partial \xi}{\partial x} + \frac{\partial \eta}{\partial y} + \frac{\partial \zeta}{\partial z} = 0. \quad (b) \qquad (16)$$

In these, ψ is a polynomial of degree n in x, y, z; and ξ, η, ζ are polynomials of degree $n-1$ in x, y, z. These polynomials are not homogeneous, but the number of coefficients in the terms of highest degree in ψ is $\frac{1}{2}(n+1)(n+2)$, and in each of ξ, η, ζ is $\frac{1}{2}n(n+1)$.

Now it has been seen that ψ and σ must be of the related forms

$$\psi = 2\pi\Sigma(H_o - H_k) A_k M_k N_k$$

and
$$\sigma = \frac{x}{a^2}\xi + \frac{y}{b^2}\eta + \frac{z}{c^2}\zeta = \Sigma A_k M_k N_k.$$

Hence if $R_k = L_k M_k N_k$ is any Lamé polynomial of order n, we have

$$\iint R_k \psi p\, dS = 2\pi(H_o - H_k) A_k L_k \iint M_k^2 N_k^2 p\, dS,$$

and
$$\iint R_k \sigma p\, dS = \qquad A_k L_k \iint M_k^2 N_k^2 p\, dS.$$

Hence, combining these results, we have the following $2n+1$ conditions on ψ, ξ, η, ζ arising from the relationship between ψ and σ

$$\int\int R_k \psi p \, dS = 2\pi(H_o - H_k) \int\int R_k \left(\frac{x}{a^2}\xi + \frac{y}{b^2}\eta + \frac{z}{c^2}\zeta\right) p \, dS, \tag{17}$$

all the integrals being taken over the surface of the ellipsoid.

As has been seen, the equation for λ, giving the frequencies of the oscillations, is to be obtained by eliminating the coefficients of the polynomials ψ, ξ, η, ζ. To obtain the frequencies *of order* n, it is necessary to effect a process of elimination on the terms of highest degree in x, y, z in the polynomials ψ, ξ, η, ζ, for it is these that contain (only) the constants associated with the MN's of order n. The lower terms will contain constants associated with surface harmonics of lower orders. It will be convenient, however, to retain the same letters (ψ, ξ, η, ζ) to denote the relevant parts of these polynomials comprising only their highest order terms.

To consider the elimination in further detail, let us write $\psi = \lambda\phi$, then equations (5) or (16) become

$$\left. \begin{aligned} \frac{\partial\phi}{\partial x} &= \lambda\xi + 2i\omega\eta, \\ \frac{\partial\phi}{\partial y} &= \lambda\eta - 2i\omega\xi, \\ \frac{\partial\phi}{\partial z} &= \lambda\zeta. \end{aligned} \right\} \ (a) \qquad \frac{\partial\xi}{\partial x} + \frac{\partial\eta}{\partial y} + \frac{\partial\zeta}{\partial z} = 0, \quad (b) \tag{18}$$

together with

$$\lambda \int\int R_k \phi p \, dS = 2\pi(H_o - H_k) \int\int R_k \left(\frac{x}{a^2}\xi + \frac{y}{b^2}\eta + \frac{z}{c^2}\zeta\right) p \, dS. \tag{19}$$

Suppose the elimination is begun by dealing first with the coefficients of the polynomial ϕ, which refers now to simply the highest terms of the original $\phi = \psi/\lambda$. If we consider the three equations (18a) involving ϕ, the coefficient of a term like x^r, or y^r, or z^r in ϕ will appear *once* only; the coefficient of a term like $x^r y^s$, or $y^r z^s$, or $z^r x^s$ will appear in *two* places; while the coefficient of a term like $x^r y^s z^t$ will appear in *three* places. For example, suppose the polynomials are denoted by:

$$\phi = a_n x^n + \ldots + a_{r,s} x^r y^s + \ldots + a_{r,s,t} x^r y^s z^t$$

$$\xi = b_{n-1} x^{n-1} + \ldots + b_{r-1,s} x^{r-1} y^s + b_{r,s-1} x^r y^{s-1} + \ldots$$
$$+ b_{r-1,s,t} x^{r-1} y^s z^t + b_{r,s-1,t} x^r y^{s-1} z^t + b_{r,s,t-1} x^r y^s z^{t-1} + \ldots.$$

$$\eta = c_{n-1} x^{n-1} + \text{etc.}$$

$$\zeta = d_{n-1} x^{n-1} + \text{etc.}$$

The equation $\quad \dfrac{\partial \phi}{\partial x} = \lambda \xi + 2i\omega\eta \quad$ must then give rise to the identity

$$na_n x^{n-1} + (n-1)a_{n-1,1} x^{n-2} y + \ldots + ra_{r,s} x^{r-1} y^s + \ldots + ra_{r,s,t} x^{r-1} y^s z^t + \ldots$$
$$= \lambda[b_{n-1} x^{n-1} + b_{n-2,1} x^{n-2} y + \ldots + b_{r-1,s} x^{r-1} y^s + \ldots + b_{r-1,s,t} x^{r-1} y^s z^t + \ldots]$$
$$+ 2i\omega[c_{n+1} x^{n-1} + \ldots].$$

Equating coefficients of the various terms gives the following necessary relations between the original coefficients:

$$
\left.
\begin{array}{lll}
x^{n-1} : & na_n = \lambda b_{n-1} + 2i\omega c_{n-1} \\
x^{n-2} y : & (n-1)a_{n-1,1} = \lambda b_{n-2,1} + 2i\omega c_{n-2,1} \\
\cdots & \cdots \qquad \cdots \qquad \cdots \\
x^{r-1} y^s : & ra_{r,s} = \lambda b_{r-1,s} + 2i\omega c_{r-1,s} \\
\cdots & \cdots \qquad \cdots \qquad \cdots \\
x^{r-1} y^s z^t : & ra_{r,s,t} = \lambda b_{r-1,s,t} + 2i\omega c_{r-1,s,t}
\end{array}
\right\}
\qquad (20)
$$

The coefficients in ϕ associated with terms such as x^r that appear once only in these equations obviously cannot be eliminated at this stage. On the other hand the coefficients in ϕ that qualify terms like $x^r y^s$ appear twice and therefore can be eliminated. The number of such coefficients is $3(n-1)$. Since the ϕ-terms are not multiplied by λ in the three equations $(18a)$, the elimination of these coefficients leads to $3(n-1)$ relations linear in λ and in the coefficients of the polynomials ξ, η, ζ.

Considering next the coefficients in ϕ that qualify terms like $x^r y^s z^t$, these appear three times each in equations $(18a)$, and the total number of such coefficients is $\frac{1}{2}(n-1)(n-2)$ [viz. $\frac{1}{2}(n+1)(n+2) - 3(n-1) - 3$]. Each of these can be eliminated in *two* independent ways, and provide two equations linear in λ and the coefficients associated with ξ, η, ζ; that is, $(n-1)(n-2)$ equations in all.

Thus the elimination in this way of those coefficients of ϕ that can be eliminated by equations $(18a)$ leads in all to

$$3(n-1) + (n-1)(n-2) = n^2 - 1$$

equations, in each of which λ and polynomial coefficients occur only linearly.

It will be noticed also that none of the coefficients of ϕ appears multiplied by λ in any of the equations derived from $(18a)$, and hence if the values of these coefficients in terms of those of ξ, η, ζ (and λ) are substituted on the left-hand side of (19), then λ will occur both as λ^2 and λ.

The total number of coefficients remaining now to be eliminated is simply the number of coefficients of ξ, η and ζ taken together, that is $\frac{3}{2}n(n+1)$. For this purpose there are already available $n^2 - 1$ equations linear in the coefficients of ξ, η, ζ and containing λ to the first power. There are also the $2n+1$ equations provided by (19) which contain λ both to the first and second powers. And finally there are the equations arising from $\mathrm{div}(\xi, \eta, \zeta) = 0$ which give $\frac{1}{2}n(n-1)$ relations not containing λ at all. The total number available is thus

$$(n^2 - 1) + (2n+1) + \tfrac{1}{2}n(n-1) = \tfrac{3}{2}n(n+1),$$

which is precisely the requisite number. The elimination of the coefficients of ξ, η, ζ can therefore be effected by means of a determinant $\Delta_n(\lambda)$ in which

    the first $n^2 - 1$ rows contain elements of order λ,

    the next $2n + 1$ rows contain elements of order λ^2,

    the remaining $\frac{1}{2}n(n-1)$ rows contain elements independent of λ.

Hence the degree in λ of the eliminant is at once

$$n^2 - 1 + 2(2n+1) = n^2 + 4n + 1.$$

Since the roots in λ of $\Delta_n(\lambda) = 0$ are equal and opposite in pairs, this number $n^2 + 4n + 1$ when even gives twice the number of independent harmonic motions possible, and when odd is twice this number plus one. Thus for $n = 3$, the total number of independent harmonic vibrations of a stable Jacobi ellipsoid is eleven.

The degree $n^2 + 4n + 1$ is always attained

The next step in the argument is to show that this degree is always reached. If in any circumstances it were not attained, this would mean that if in the set of equations finally giving rise to the determinant $\Delta_n(\lambda)$ only the terms of highest degree in λ in each element were retained, the resulting set of equations would be linearly dependent.

It is to be remembered that those coefficients of λ that can be eliminated are first removed to provide $3(n-1)$ relations between the coefficients of the ξ, η, ζ polynomials, and these are some of the equations in which now only terms in λ are being retained. Writing for the moment $\lambda\chi = \phi$, then first of all the coefficients of χ are eliminated. Thus, the linear dependence, after retaining only the highest terms in λ, is equivalent to asserting the consistency of the following set of equations

$$\left.\begin{array}{c} \dfrac{\partial \chi}{\partial x} = \xi, \quad \dfrac{\partial \chi}{\partial y} = \eta, \quad \dfrac{\partial \chi}{\partial z} = \zeta, \\[2mm] \dfrac{\partial \xi}{\partial x} + \dfrac{\partial \eta}{\partial y} + \dfrac{\partial \zeta}{\partial z} = 0, \quad \text{i.e. } \nabla^2\chi = 0. \\[2mm] \text{and} \qquad \displaystyle\int\!\!\int R_k \chi p \, dS = 0. \end{array}\right\} \tag{21}$$

But if this set of equations held, it would mean that χ is a harmonic polynomial of degree n orthogonal to all the Lamé polynomials of degree n, and this cannot be so.

Hence it follows that the degree of the highest power of λ occurring in $\Delta_n(\lambda)$ is essentially $n^2 + 4n + 1$.

Form of the constant term in $\Delta_n(\lambda)$

This term will be obtained simply by putting $\lambda = 0$ in every element of the determinant. Now if in the $2n + 1$ rows that arise from (17) we put $\lambda = 0$, the remaining terms have a common factor of $H_o - H_k$. For putting $\lambda = 0$, (17) becomes

$$0 = 2\pi(H_o - H_k)\iint R_k\left(\frac{x}{a^2}\xi + \frac{y}{b^2}\eta + \frac{z}{c^2}\zeta\right)p\,dS,$$

and the right-hand side is linear in the coefficients of ξ, η, ζ and also independent of λ. Hence in $\Delta_n(\lambda)$ the constant term contains each of the coefficients of stability $H_o - H_k$ as factor, and thus $2n+1$ such factors in all. We must therefore have in the first place

$$\Delta_n(0) = M \prod_1^{2n+1} (H_o - H_k),$$

where M is some multiplying factor, to be considered next. This form is what would be expected by analogy with systems of a finite number of degrees of freedom (see p. 20).

We proceed to consider now whether M vanishes in any particular circumstances. If M were zero, it would imply the consistency of the equations got by putting $\lambda = 0$ in (16) and (17). That is, the following set of equations would be self-consistent:

$$\left.\begin{aligned}
&\frac{\partial\phi}{\partial x} = 2i\omega\eta, \quad \frac{\partial\phi}{\partial y} = -2i\omega\xi, \quad \frac{\partial\phi}{\partial z} = 0,\\[2mm]
&\frac{\partial\xi}{\partial x} + \frac{\partial\eta}{\partial y} + \frac{\partial\zeta}{\partial z} = 0,\\[2mm]
&0 = \iint R_k\left(\frac{x}{a^2}\xi + \frac{y}{b^2}\eta + \frac{z}{c^2}\zeta\right)p\,dS \quad (k = 1, 2, ..., 2n+1).
\end{aligned}\right\} \tag{22}$$

Assuming this to be so, it is simple to find what the equations imply. For, since $\frac{\partial\phi}{\partial z} = 0$, ϕ must be independent of z. Hence ξ, η are also independent of z, since they are proportional to derivatives of ϕ. Also we then have

$$\frac{\partial\zeta}{\partial z} = -\frac{\partial\xi}{\partial x} - \frac{\partial\eta}{\partial y} = \frac{1}{2i\omega}\left(\frac{\partial^2\phi}{\partial x\,\partial y} - \frac{\partial^2\phi}{\partial y\,\partial x}\right) = 0.$$

Hence ζ is independent of z. It follows then that the expression

$$\frac{x}{a^2}\xi + \frac{y}{b^2}\eta + \frac{z}{c^2}\zeta \equiv -\frac{1}{2i\omega}\left(\frac{x}{a^2}\frac{\partial\phi}{\partial y} - \frac{y}{b^2}\frac{\partial\phi}{\partial z}\right) + \frac{\zeta}{c^2}z$$

is a polynomial of degree n containing z only to the first power, where it occurs explicitly in the last term. But then the last of equations (22) means that this polynomial (linear in z) reduces on the surface of the ellipsoid to a sum of Lamé polynomials of order less than n. But the degree of $\frac{x}{a^2}\xi + \frac{y}{b^2}\eta + \frac{z}{c^2}\zeta$ is necessarily n, since ξ, η, ζ are each of degree $n-1$. Hence the situation must be that

$$\frac{x}{a^2}\xi + \frac{y}{b^2}\eta + \frac{z}{c^2}\zeta = \text{(polynomial of degree } n-2)\left(\frac{x^2}{a^2} + \frac{y^2}{b^2} + \frac{z^2}{c^2} - 1\right)$$

$$+ \text{polynomial of degree } n-1 \text{ or less.}$$

But ξ, η, ζ now represent homogeneous polynomials (since we are retaining only terms of order n in finding $\Delta_n(\lambda)$), and hence $\frac{x}{a^2}\xi + \frac{y}{b^2}\eta + \frac{z}{c^2}\zeta$ is also homogeneous of degree n. If therefore we rewrite the above in the form

$$\frac{x}{a^2}\xi + \frac{y}{b^2}\eta + \frac{z}{c^2}\zeta = \text{(homogeneous polynomial of degree } n-2\text{)} \left(\frac{x^2}{a^2} + \frac{y^2}{b^2} + \frac{z^2}{c^2}\right)$$

$$+ \textit{polynomial of degree } n\textit{–}1 \textit{ or less},$$

this must be homogeneous of degree n, and hence the last term, in italics, must be zero. Thus $\frac{x}{a^2}\xi + \frac{y}{b^2}\eta + \frac{z}{c^2}\zeta$ must be divisible by $\frac{x^2}{a^2} + \frac{y^2}{b^2} + \frac{z^2}{c^2}$, which is impossible (since it contains z only to the first power) unless it is identically 0. This requires first $\zeta = 0$, since the term in z must vanish identically, and second

$$\frac{x}{a^2}\frac{\partial \phi}{\partial y} = \frac{y}{b^2}\frac{\partial \phi}{\partial x}.$$

This second requirement means that ϕ depends solely on $\frac{x^2}{a^2} + \frac{y^2}{b^2}$, and since it is homogeneous it must be proportional simply to some power of this; moreover, since it is of degree n, we must have

$$\phi = \left(\frac{x^2}{a^2} + \frac{y^2}{b^2}\right)^{\frac{1}{2}n},$$

and this is only a rational polynomial if n is even.

It may be concluded, therefore, that the equations (22) *are* self-consistent and that M vanishes when n is even.

While if n is odd, M is essentially non-zero.

It would be expected that $M = 0$ for n even, since then the total number of roots $n^2 + 4n + 1$ is odd, and since the roots are equal and opposite in pairs $\lambda = 0$ must be a root. The case of special importance to our purpose is $n = 3$, and the constant term is then

$$\Delta_n(0) = M \text{ (non-zero)} \prod_1^{2n+1} (H_o - H_k).$$

(Note: When in the equations resulting from (18) and (19) we retain only the terms in λ, or later only the terms not containing λ, the function ϕ then appearing denotes of course only that part of the original ϕ balancing the retained terms on the right.)

Ordinary stability ceases simultaneously with secular stability

We can now establish the important result that the Jacobi ellipsoid becomes ordinarily unstable simultaneously with its becoming secularly unstable through the changing of sign of the characteristic coefficient of stability of order three.

For if we denote C_n, the characteristic coefficient of stability of order n, by $H_o - H_{2n+1}$, then for $C_n > 0$, while the ellipsoid is thoroughly stable, we have by the foregoing result

$$\lambda_1^2 \lambda_2^2 \ldots \lambda_m^2 = \theta(H_o - H_{2n+1}) > 0, \tag{23}$$

where we may suppose $\lambda_1 < \lambda_2 < \ldots < \lambda_m$ are the positive roots of $\Delta_n(\lambda) = 0$, and are m in number where $2m = n^2 + 4n + 1$. Also θ is the product of M with the remaining non-vanishing coefficients of stability, and must therefore always remain positive since it is positive initially. On the other hand, for a Jacobi figure just beyond the critical ellipsoid, we must have

$$\lambda_1^2 \lambda_2^2 \ldots \lambda_m^2 = \theta(H_o - H_{2n+1}) < 0. \tag{24}$$

Now, by continuity, only the smallest of the roots, viz. λ_1^2, can have changed sign, and hence now λ_1^2 must be negative, λ_1 therefore imaginary, and the system *ordinarily unstable*.

It is not necessary to consider specially the possibility of the coincidence of the smallest roots of $\lambda_1^2 = \lambda_2^2$, for then we should have simply λ_1^4 negative, and so λ_1 would be complex, and exponential terms with real indices would enter just the same.

Finally, it may be noted (cf. p. 84) that for a third-order harmonic deformation the angular velocity is not changed to the first order. Hence the instability is a true one not capable of being removed by selecting a frame of reference with a slightly different rate of rotation.

This completes the proof of the ordinary instability of the Jacobi ellipsoids beyond the point at which the pear-shaped series bifurcates.

Chapter X

COSMOGONICAL IMPLICATIONS

Impossibility of the fission hypothesis

According to the fission theory, the origin of binary stars is regarded as due to the break-up of a single mass by its rotation. There are a number of grave astrophysical difficulties in this theory such as the question of the source of the necessary angular momentum, whether the density distribution in stars and its evolution are consistent with fissional disruption, whether the distribution of mass-ratios and separations are explicable on the theory, and the whole general problem of stellar evolution itself. We will not attempt to discuss these here but will concern ourselves solely with examining the fission process itself, as proposed by its advocates, on purely dynamical grounds. A general secular increase of angular momentum will be assumed, with the density remaining uniform, since this is equivalent to constant angular momentum with density increasing gradually. These are the assumptions on which both Darwin and Jeans based their ideas on fission.

Now, as already explained in the introduction, had Darwin's conclusion, that the pear-shaped figure is secularly stable, been correct, it might then have been fairly plausible to suppose that the deepening of the furrow with evolution along the series was some indication that the mass would eventually divide into two parts in orbital motion about each other (though information on the initial stages of this process by no means necessarily secured that the pear-shaped series itself did not bifurcate later to some new form). But when Jeans's studies, in agreement with those of Liapounoff, contradicted Darwin's conclusion, the sole grounds on which Darwin based his description of the fissional process were completely removed. Yet Jeans nevertheless finally maintained exactly the same outcome of the process as Darwin, namely, fission into two detached masses moving in almost circular orbits about each other. Jeans's view obviously amounted to asserting that the evolution of the mass would be quite independent of whether the pear-shaped series happened to be secularly stable or unstable, and hence that the whole of the investigations establishing the incorrectness of Darwin's conclusions, or indeed any studies of this problem at all, were valueless from the point of view of cosmogony, since the outcome of the process in either case would be the same. As the fission theory will always be associated with Jeans, and as its acceptance is still one of the main errors besetting even modern writings on stellar evolution, it seems worth while dealing in some detail with the many points at which Jeans fell into error.

The first of these concerns the nature of the difference between secular and ordinary stability of which at many parts of his work Jeans seems to have had an inadequate conception. Thus at one point* he says:

* *Astronomy and Cosmogony*, p. 199.

. . . as the physical conditions of a system gradually change secular stability necessarily sets in *before* ordinary stability. Thus, for problems of cosmogony it is secular
instability alone which is of interest. A system never attains to a configuration in
which ordinary instability comes into operation since secular instability must always
have *previously* intervened.

Now this statement is plainly incorrect, since, although a system cannot become
ordinarily unstable before becoming secularly unstable, the two kinds of stability
can vanish simultaneously. Any finite system for which a single coefficient of
stability changes sign, for example, will become both ordinarily and secularly
unstable at the same stage, since the product of the squares of the frequencies
of oscillations has always the same sign as the product of the coefficients of
stability. A simple instance of this was discussed in Chapter II, where it was
seen that a particle in a circular orbit under a central force μ/r^n becomes simultaneously both secularly and ordinarily unstable when n rises above 3. Also we
have seen that the same actually holds for the Jacobian ellipsoids, which become
both secularly and ordinarily unstable for third-order harmonic displacements at
the same stage. Naturally enough Jeans, holding the view he stated, failed to
see the possible importance of ordinary instability. But as it happens, it can be
shown on general grounds, from the property that the ellipsoids become ordinarily
unstable, that the fission process as described by Jeans is dynamically impossible.
For the essence of ordinary instability is that it is independent of friction, and
therefore that any motion ensuing from it must be of a strictly reversible character; that is, if $-t$ is written for t in the equations of motion they must remain
unaltered. But if in the complete absence of friction we imagine a close double-
system as the product of rotational instability of a single ellipsoidal mass, a
reversal of the motion of the bodies simply leads to them describing their orbits
in the reverse direction and not to their re-uniting into the original mass. Thus
at a single blow the fact that the system becomes ordinarily unstable disposes
of the fission hypothesis for the formation of double systems.

But Jeans seems even to have misunderstood the nature of secular instability
too. It is clear, from the fact that the initial members of the pear-shaped series
have *less* angular momentum than the last stable ellipsoid, that the system cannot
evolve along the pear-shaped series, since this would necessitate a continually
decreasing angular momentum for the body. Yet Jeans supposed it to be possible
to find the initial details of the process of break-up by fission by calculating to a
fairly high order of accuracy the shapes of the members of the pear-shaped
series in the corresponding two-dimensional cylindrical problem which has
analogous properties to the three-dimensional problem but is far simpler of
treatment. The resulting figures (*Problems of Cosmogony*, p. 116; *Astronomy and
Cosmogony*, pp. 220, 221) in fact can establish nothing of the precise course of
development when the critical Jacobi form is passed. For the ordinary instability
at this stage means that the precise further development of the system depends
on the detailed initial conditions of disturbance, and these from their nature are
essentially unknowable and different from one system to another, though it may
nevertheless happen that the general outcome of the instability is independent
of the exact nature of the disturbance. Thus, a rod standing upright on a fixed

pivot at its lower end and acted on solely by gravity is ordinarily unstable, but the plane in which it falls away from this position and proceeds to move like a pendulum depends entirely on the initial disturbance given to it, however small this may be. Again, a particle describing an unstable circular orbit under a central force, when disturbed may either eventually fall into the centre of force or depart an infinite distance from it. There are only the two possible general results, but the paths by which the particle may reach either one of them are infinite in number and depend on the precise conditions of disturbance. Thus the final outcome of the instability of the Jacobi form could conceivably be fairly definite though the intervening motion might be subject to complete uncertainty.

In some parts of his writings on the fission problem, Jeans seems waveringly to have held various opinions quite inconsistent with others expressed elsewhere in his work. Thus he says:*

> The fact that the pear-shaped series is initially unstable shews that a rotating mass cannot evolve by slow secular changes through a series of pear-shaped figures. This somewhat diminishes the interest of the pear-shaped series in the problem of cosmogony, but nevertheless it remains important to obtain as clear an idea as we can of the nature of the series.

Yet a few pages later,† after having calculated the two-dimensional forms, as explained above, he says:

> Thus we may with fair confidence assert that the two-dimensional [pear-shaped] series ends by fission into two detached masses and . . . it seems highly probable that the three-dimensional series also will end by a similar fission.

At a later point in the same work‡ Jeans appears to have still another view of the implications of secular instability, for he there says:

> The pear-shaped figure is unstable, so that as soon as it is formed dynamical motion ensues and fission results. The masses are at first projected away from one another with considerable velocity, but seem likely to settle down finally to describe steady orbits about one another.
> In his researches on this problem, Darwin supposed that the initial orbits would be strictly circular, but this was because he believed the process of fission to be a statical process and not a dynamical process, as we have seen it to be. According to Darwin's view, the changes in the star while fission was taking place were, initially at least, of a purely secular nature, and it was natural to suppose that the final result would be two masses rotating in actual contact and at rest relative to one another.
> There being no longer any theoretical justification for supposing that the initial orbits will be strictly circular, we have to consider the possibility of the masses being thrown apart with appreciable radial velocities, and describing elliptic orbits about one another.
> Consider for simplicity the case in which the original star is supposed to divide into equal masses, and suppose that fission occurs when the centre of each mass is at a distance r from the common centre of gravity. Let each star be supposed to have a radial velocity v in addition to the tangential velocity ωr in space resulting from rotation. Each mass will describe approximately an elliptic orbit in space so that after the orbits are partially described the masses will again each be at a distance r from their common centre of gravity, but are now approaching each other with a radial velocity v. A collision of some kind must occur, and since the masses will not

---

* *Problems of Cosmogony*, p. 102.

† *Ibid.*, p. 115.

‡ *Ibid.*, p. 252.

be perfectly elastic, their velocity of recession after collision will be some velocity v_0 less than v, while the [transverse]* velocity ωr must, from the conservation of angular momentum, be the same as before. It follows that the new orbit will be of less eccentricity than the old, and the eccentricity will further diminish at each subsequent collision. We cannot argue that the eccentricity will be finally reduced to zero; a limiting value will be reached such that the masses just graze one another at periastron.

This account of the matter shows some appreciation of the likely consequences of instability apart from overlooking the possibility that v might be great enough for escape to occur, but even so there are a number of other strong objections to the suggested course of development. To begin with, the assumption of equal masses after break-up can easily be shown to be invalid since such a system would necessarily require more angular momentum than the last stable Jacobi form, but is of small importance compared with the ideas expressed concerning the collisions. If the masses are projected away from each other with considerable velocity, as may reasonably be assumed, they will perforce come together with practically the same velocity, as stated, but the two masses would re-unite— colliding liquids do not retain their identities and separate like billiard balls— and there seems no possibility of them retaining their individual identity in the way Jeans supposes. But even if this difficulty is passed over, there is the objection that the angular momentum of the critical Jacobi form is appreciably smaller than that of two equal or comparable masses in stable orbital motion about each other (see Table VI, p. 142), and the discrepancy would be increased by the introduction of any eccentricity for the purpose of keeping the smaller mass outside the limit within which it is tidally unstable.

Although these ideas of Jeans are somewhat confused it may nevertheless be the case that when the Jacobi form becomes ordinarily unstable a slight disturbance will cause it eventually to divide into two masses. For there is little doubt that some sort of break-up must occur for the reason that the single system possesses more angular momentum than it can store as a stable mass in any known form. By some means the mass must rid itself of at least part of the angular momentum, and its transference to that of orbital motion of separate pieces is obviously a possible way of effecting this. No other way out of the difficulty has ever been proposed apart from ejection of a part of the mass and no other way seems physically possible. But even so it remains necessary to avoid the difficulty arrived at above concerning the collision and re-uniting of the pieces. If, for simplicity of discussion, we assume break-up into two masses, the only way of avoiding a subsequent coalescence of the pieces is if the initial rate of separation is sufficiently great to carry the pieces to infinity, that is, for them to have hyperbolic relative speed of separation. If this did not happen at first, the collision and re-uniting of the masses would necessarily involve dissipation of energy and a rendering of the system more unstable still. (The ultimate effect of dissipation owing to a collision would be an increase of density and this would be equivalent to an increase of angular momentum without change of density.) A further break-up, but this time with greater violence would follow,

* Jeans has 'radial'.

and so on, until a disruption was reached that succeeded in sending the pieces apart with hyperbolic speed. It appears that in no other manner can the system find its way to a new state of stable steady motion. Each piece will then become a separate stable rotating system and the relative velocity will eventually decrease to a constant value.

It does not seem to be generally known, and it is certainly not made clear in either of Jeans's books (*Problems of Cosmogony*, 1919; *Astronomy and Cosmogony*, 1929), that the conclusion originally arrived at by him concerning the stability of the two-dimensional pear-shaped figures was incorrect. Thus in 1902* Jeans announced that his analysis established that the cylindrical pear-shaped series was secularly stable, a conclusion that seemed to lend strong support to Darwin's corresponding conclusion for the three-dimensional series. It was then natural to suppose that further evolution would take place, at least to begin with, along the pear-shaped series with a gradual deepening of the furrow, suggesting eventual division into two masses, and in the two-dimensional case the forms could be calculated to a fairly high order of accuracy. But many years later in 1916† Jeans briefly mentions that his former value for the moment of inertia of the two-dimensional pear was in fact wrong owing to a simple numerical error and that the corrected value implied instability. This signal discovery appears to pass unmentioned in both the 1919 and 1929 books, though the unstable nature of the forms is concluded there, but curiously enough Jeans retains the remaining conclusions of his 1902 paper in detail and again claims, as then, that the mass will still evolve along the unstable series.

To sum up on the question of the fission theory of the origin of binary stars: even if all its extremely doubtful initial assumptions are permitted to give the theory its most favourable conditions for success, we are led purely by dynamical evidence to the conclusion that the process of instability cannot result in the formation of a double system consisting of two comparable masses moving in close orbital motion about each other. When it is remembered that added to this the particular theory of stellar evolution of which fission forms a crucial part itself depends to a large extent on conjectural speculations on the internal structure and source of energy of the stars, long since known to be invalid, there appears to be no remaining evidence in favour of the fission theory.

The origin of satellites within the solar system

It does not of course follow that the process of break-up with ejection of part of the mass to infinity has played no part at all in cosmogonical changes in any celestial systems, but if the present dynamical theory is to be applicable to even a moderately close degree, it can only be for systems having some fair approximation to uniform density in the greater portion of their mass. There is reason to believe that this holds for the main bodies of the planets of the solar system, though for reasons that will become clear we shall not attempt to apply it directly to any of the planets in their present forms but instead only to what may be

* *Phil. Trans.* 200 A, p. 96.
† *Ibid.* 217 A, p. 28.

termed primitive planets, by which is meant the forms that these bodies may have taken at an earlier stage in their development. There are cogent physical arguments for supposing that in its early stages the solar system developed through a state that consisted of a number of bodies having planetary distances from the sun and arriving at a condition of rotational instability either through increase of density or through the addition at their surfaces of matter endowed with the necessary angular momentum, or through possibly a combination of both processes.

According to certain hypotheses of the origin of the solar system, the planetary material was initially detached from that of a companion star to the sun moving at such a separation for its material to have angular momentum per unit mass comparable with that at present associated with the great outer planets. It is not necessary for our present purpose to make any decision between the various hypotheses of this kind, for in all of them the result is arrived at that only a few large planets, comparable in mass at least with those of the present great planets, could be formed initially. The problem then remains how the numerous satellite systems have arisen. The satellites are of such small masses that there is no possibility of them forming directly by condensation from material at stellar temperatures, nor could they form and remain in orbital motion round a planet itself forming by condensation and contraction from a disc of material revolving round the sun. Yet another difficulty is the existence of the four much smaller terrestrial planets which exhibit surprising differences of composition among themselves. Of these planets Mercury in particular, and Mars to some extent, are more nearly comparable in mass with the largest satellites than with other planets.

The problem can be approached from a slightly different point of view by considering the general question of the contraction through gravitational causes of a mass of material endowed with angular momentum. With continually increasing density a state of rotational instability must always arise sooner or later unless some change of physical state enters the system to arrest further increase of density. In most stars, for instance, energy generation sets in in the central regions long before the density increase becomes consistent with rotational instability of any kind. In a planet what evidently finally protects the system from rotational instability is its reaching a liquid or solid state such that further contraction ceases. But there is no reason to expect that the angular momentum per unit mass associated with the material from which any primitive planet formed would be insufficient to cause instability before such a state was reached. If the rotational momentum were great enough, a primitive planet would become unstable before this and result in a fissional break-up with the ejection to infinity of a part of the mass. For uniform density it can be shown in the following way that the portion so lost to the main body must be less than about one-eighth of the original mass.

Suppose the original mass M breaks into two separate pieces of individual masses mM and $(1-m)M$, where $m < \frac{1}{2}$, without change of density, and which eventually just escape from each other to infinity. The speeds of rotation of these masses when they are first formed would probably not differ much from that of the critical Jacobi ellipsoid, but their shapes will not be critical ellipsoids. They could, in fact, possess this particular angular velocity (given by $\omega^2/2\pi G\rho = 0\cdot142$) while having the form of a thoroughly stable Maclaurin spheroid of eccentricity of section about $0\cdot71$, as is shown by Table I, p. 39. The masses could, however, retain most energy when in the critical Jacobi form, and if we attempt to impose this upper limit of internal energy on the dividing pieces, the condition they can just succeed in escaping from each other is, at once

$$-0\cdot745\,GM^{5/3}\rho^{1/3} \geqslant -0\cdot745\,G(mM)^{5/3}\rho^{1/3} - 0\cdot745\,G(1-m)^{5/3}M^{5/3}\rho^{1/3},$$

or
$$1 \leqslant m^{5/3} + (1-m)^{5/3},$$

and this cannot be satisfied for any non-zero m in the range $0 < m < \frac{1}{2}$. This establishes, therefore, that if the pieces into which the original mass breaks eventually separate to infinity, they will do so as stable bodies (excluding, of course, the possibility of further internal contraction).

If at the other extreme it is assumed that the two resulting components eventually settle down into bodies of negligible rotation, so that they retain the least possible amount of energy, the internal energies of the pieces will be given by the appropriate expression for the spherical form (Chapter IV, equation 6), and the energy condition for complete separation of the components is now, omitting common factors,

$$-0\cdot745 \geqslant -0\cdot967\,m^{5/3} - 0\cdot967\,(1-m)^{5/3},$$

or
$$0\cdot77 \leqslant m^{5/3} + (1-m)^{5/3},$$

and this requires $m < 0\cdot19$, giving a limiting mass-ratio of rather more than $4:1$. But in any actual case the mass-ratio must *exceed* this value, for the components must inevitably possess a certain amount of rotation.

If the reasonable supposition is made that the angular velocity of the components is the same as that of the critical ellipsoid, but that the separate pieces have the form of Maclaurin spheroids, then it is readily found by interpolation from Table I (p. 39) that when $e = 0\cdot71$, the potential and kinetic energies for a mass mM are given by

$$V = -0\cdot9571\,G(mM)^{5/3}\rho^{1/3}$$

$$T = 0\cdot0868\,G(mM)^{5/3}\rho^{1/3}$$

in the units of the present discussion. Accordingly, the total energy of the component mM when separated to infinite distance from $(1-m)M$ is $-0\cdot8703\,G(mM)^{5/3}\rho^{1/3}$, and the energy condition in this case is

$$-0\cdot745 \geqslant -0\cdot8703\{m^{5/3} + (1-m)^{5/3}\},$$

or
$$0\cdot856 \leqslant m^{5/3} + (1-m)^{5/3}.$$

This form of the condition requires $m < 0\cdot11$ about, and the limiting mass-ratio is therefore about $8:1$.

Turning now to the question of the conservation of angular momentum during the disruption, this must similarly place a restriction on m. But a slight difficulty is met with here in that the orbital angular momentum of the separating pieces is determined solely by the transverse component of relative velocity of the individual masses, whereas the criterion for escape involves only the total relative velocity and its direction is immaterial. If it is assumed that the transverse relative velocity of the smaller mass itself involves sufficient energy for escape, it is readily found that the parabolic orbital angular momentum of mM and $(1-m)M$, supposed to possess spherical form in making this part of the calculation, is given by

$$1 \cdot 12 \, G^{1/2} M^{5/3} \rho^{-1/6} m(1-m) \sqrt{[m^{1/3} + (1-m)^{1/3}]}.$$

Also, the angular momentum of the critical ellipsoid is

$$0 \cdot 31 \, G^{1/2} M^{5/3} \rho^{-1/6}.$$

If now, as in considering the energy balance, it is supposed that mM and $(1-m)M$ reach great separation in the form of Maclaurin spheroids with $\omega^2/2\pi G\rho = 0 \cdot 142$, then it is readily found from Table I (p. 39), that in the present units the angular momentum of mM is given by

$$0 \cdot 18 \, G^{1/2} (mM)^{5/3} \rho^{-1/6},$$

with an exactly similar expression for $(1-m)M$. Accordingly the conservation of angular momentum requires

$$0 \cdot 31 \geqslant 0 \cdot 18 \, m^{5/3} + 0 \cdot 18 \, (1-m)^{5/3} + 1 \cdot 12 \, m(1-m) \sqrt{[m^{1/3} + (1-m)^{1/3}]},$$

or $\qquad 1 \cdot 69 \geqslant m^{5/3} + (1-m)^{5/3} + 6 \cdot 10 \, m(1-m) \sqrt{[m^{1/3} + (1-m)^{1/3}]},$

and this requires a value of m at most slightly less than $0 \cdot 14$, in other words, a mass-ratio rather greater than 7 : 1.

In view of the unavoidable element of approximation involved in the foregoing calculations, the general agreement between the two limits, arising separately from the energy and angular momentum requirements, gives strong indication that a mass-ratio of at least 7 or 8 to 1 must be the outcome of the instability.

Consideration of double systems in circular orbits

The validity or otherwise of the fission mechanism for the formation of double systems can be investigated from another standpoint by considering the dynamical properties of stable double systems, and it is found that this provides further evidence on the question of the mass-ratio and the ultimate result of fission.

The appropriate dynamical problem has been studied by Darwin, who has shown that if two masses of equal uniform density are moving in circular orbits about each other so that the whole motion is effectively one of rigid-body rotation, the forms of the free surfaces of the components are to a high degree of approximation representable by ellipsoids. If we adopt Darwin's notation and denote the masses by m and M, then $\lambda = m/M$ will be the mass-ratio. The unit of length may also be chosen in such a way that

$$m = \frac{4}{3}\pi\rho \cdot \frac{\lambda}{1+\lambda}, \quad M = \frac{4}{3}\pi\rho \cdot \frac{1}{1+\lambda},$$

so that the total combined mass, without change of density, would occupy a sphere of unit radius. The distance between the centres of the two masses is denoted by R, and the axes of m by a, b, c, and those of M by A, B, C. Then Darwin's calculations show that in relative stable equilibrium the two ellipsoids are so oriented that the longest axes, denoted now by c and C, are in the line joining the centres of mass, while the two axes of intermediate length, b and B, are also in the plane of the orbital motion. The remaining axes, a and A, are accordingly in directions parallel to the axis of rotation of the system as a whole.

The problem studied by Darwin was to find the limiting least value of the separation of the components, denoted by R, consistent with their remaining stable configurations. In every case it is found, as might be expected, that the smaller mass, m, becomes unstable first as the distance apart decreases. The actual cases investigated by Darwin were for $\lambda = 0.4$ to 1.0 at intervals of 0.1, while the value $\lambda = 0$ corresponds to Roche's problem of an infinitesimal satellite in the presence of a primary affected only by its rotation, which is equal to that of the satellite in its orbit. Since the values of λ that turn out to be of chief interest for our present purpose are slightly less than 0.4, the calculations for $\lambda = 0.1$, 0.2, and 0.3 have been carried out to complete Darwin's discussion and generally exhibit the continuous character of the series of configurations as λ varies.

Table V shows the values of the dimensions of the system in the configuration of limiting stability.

TABLE V

λ	R	$m = \lambda M$			$M = m/\lambda$		
		a	b	c	A	B	C
0·0	2·457	$\dfrac{0.482}{\infty}$	$\dfrac{0.511}{\infty}$	$\dfrac{1.000}{\infty}$	0·942	1·030	1·030
0·1	2·465	0·363	0·387	0·647	0·905	0·987	1·018
0·2	2·471	0·456	0·486	0·752	0·872	0·949	1·007
0·3	2·477	0·517	0·553	0·807	0·842	0·916	0·997
0·4	2·484	0·562	0·603	0·843	0·815	0·886	0·988
0·5	2·485	0·597	0·642	0·870	0·792	0·860	0·979
0·6	2·490	0·627	0·674	0·888	0·772	0·836	0·969
0·7	2·497	0·654	0·701	0·901	0·753	0·815	0·958
0·8	2·502	0·673	0·725	0·912	0·737	0·796	0·947
0·9	2·508	0·691	0·744	0·921	0·722	0·778	0·937
1·0	2·514	0·708	0·762	0·927	0·708	0·762	0·927

It is obviously not necessary to consider values of λ greater than 1, since m would then simply become the larger mass and only an interchange of m and M would be needed for the purpose of investigating the stability of M for $\lambda > 1$. For $\lambda = 0$ the mass m is infinitesimal, but its shape, which is what determines its stability, is indicated by giving the ratio of its axes in limiting form.

By means of Table V we can now draw up a further table of relevant quantities. First, the quantity $R - (c + C)$ will be the shortest distance between the surfaces of the components and is of obvious importance in considering the geometrical possibility of the orbital motion. Second, may be tabulated the complementary quantity $R + (c + C)$, which measures the overall extent of the binary system, in order to compare it with the greatest extension of the critical Jacobian ellipsoid having the same volume as the combined volume of the components. The fourth and fifth columns in Table VI give the angular momentum h and the angular velocity in the form $\omega^2/2\pi G\rho$ in each of the limiting configurations. Below these are given the corresponding quantities for the critical Jacobi ellipsoid, while in the last line are given the interpolated dimensions of the double-system having the same angular momentum as the critical Jacobi figure, which corresponds to the value $\lambda = 0.36$ approximately.

TABLE VI

λ	$R-(c+C)$	$R+(c+C)$	h	$\omega^2/2\pi G\rho$
0.0	1.427	3.487	0.110	0.0449
0.1	0.800	4.130	0.227	0.0444
0.2	0.712	4.230	0.307	0.0441
0.3	0.673	4.281	0.363	0.0438
0.4	0.653	4.315	0.402	0.0435
0.5	0.636	4.334	0.428	0.0434
0.6	0.633	4.347	0.446	0.0432
0.7	0.638	4.356	0.457	0.0428
0.8	0.643	4.361	0.464	0.0426
0.9	0.650	4.366	0.468	0.0423
1.0	0.660	4.368	0.469	0.0420
Jacobi ellipsoid	twice longest axis = 3.772		0.389	0.1420
$\lambda = 0.36$	0.660	4.303	0.389	0.0436

It is seen from Table VI that for all values of λ greater than about 0.36 the binary system will have greater angular momentum than could be stored by their combined mass in the form of a single body just on the verge of rotational instability. A larger orbit involving greater stability for the smaller mass would require even more angular momentum than the critical configuration, that is, a smaller value of λ would be involved and hence a higher mass-ratio still. It is to be noticed that for no value of λ does the limiting position of stability of the smaller mass allow the two components to be nearly in contact, and in fact the overall extent of the double system (4.303) is considerably greater than the overall length of the critical Jacobian figure (3.772), so that there can be no possibility of the latter configuration evolving *gradually* into the former.

Finally may be noticed the considerable disparity between the angular velocities of the Jacobian figure and the critical double system of the same angular momentum, or indeed of any more stable double system of the same total mass. For the Jacobian figure $\omega^2/2\pi G\rho = 0.142$, whereas for $\lambda = 0.36$ the orbital angular velocity is given by $\omega^2/2\pi G\rho = 0.0436$ and is little different from this for any other value of λ. If the separation of the components exceeded the limiting value R, the corresponding value of ω would clearly be still smaller. There is therefore a difference of angular velocity between the critical Jacobian form and the double system measured by a factor of about 1·8, the former rotating much the faster. Now it is well known that if the angular velocity of two particles in a circular orbit is increased by a factor $\sqrt{2}\,(=1\cdot414)$, the particles would just escape from each other, so that there is further strong evidence here in this factor of 1·8 that the outcome of fission must be to endow the resulting components with a relative velocity sufficient for complete escape from each other, and moreover that the necessary energy may well be carried almost entirely by the initial transverse component of relative velocity, as assumed above in considering the angular momentum balance.

The formation of satellites

In accordance with the foregoing dynamical arguments it seems safe to conclude that the smaller portion must escape entirely from the gravitational field of the larger remaining mass, and the final relative velocity will be shared between the two parts in the ratio of their masses, so that relative to their common centre of mass the smaller part carries almost all the relative motion. It is suggested that the great planets represent the surviving main portions of such rotational disruptions of a number of larger primitive planets. The velocities likely to be involved can be estimated in order of magnitude by values for the actual planets. At the surface of Jupiter the escape velocity is about 60 km. sec.$^{-1}$ whereas the orbital velocity of Jupiter round the sun is about 13 km. sec.$^{-1}$ and the escape velocity from the sun at that distance about 18·5 km. sec.$^{-1}$. Thus a particle projected from Jupiter at 63 km. sec.$^{-1}$ would have sufficient energy to escape not only from the field of the planet but also from the solar system. Similarly, for Neptune the escape velocity at its surface is 23 km. sec.$^{-1}$ while the orbital velocity is 5·43 km. sec.$^{-1}$ and the parabolic speed 7·65 km. sec.$^{-1}$, so that a particle projected from Neptune at 24·3 km. sec.$^{-1}$ or higher would escape entirely. Thus the additional energy required for escape from the solar system as well as from the planet is such a small fraction of the energy of parabolic or hyperbolic ejection that the smaller component resulting from a rotational break-up will usually escape entirely from both the remaining mass and the sun, though there is clearly a small probability that this might not happen in any particular case. On the other hand, the recoil velocity of the larger component will be so much smaller that the disruption is not likely to endow it with a velocity approaching a hyperbolic value, at any rate at distances from the sun comparable with those of the great planets.

But considering the disruption itself in more detail, it is not to be expected that this would take place simply as a clear division into two separate parts, for the material of the body near where the main break occurred would necessarily be situated in the neighbourhood of a neutral dynamical point from which the main masses would be receding in opposite directions. It seems likely that a stream of material would form stretched between the masses and having velocity ranging between those of the main masses. Such a stream would not be a stable arrangement and as the main masses separated would break up into small masses under its internal gravitation. The ones near the centre of the line of small bodies would escape from both the main bodies, while some of those nearer the ends might have insufficient speeds to escape the adjacent main mass, but the general rotation associated with the original mass would of course be maintained throughout and endow each of these small masses with orbital angular momentum about it. In this way the larger residual portion might come to possess a system of small satellite masses moving about it at distances comparable with the size of the body itself and in the same general direction as that determined by the rotation of the main body.

The details of such a process inherently preclude mathematical analysis, for the removal of material from one gravitating body by the action of another, and with the addition here of a rotational field, necessarily involves motion at and near neutral points, and in such circumstances arbitrarily small changes in conditions can produce arbitrarily large differences in the resulting motions. This mathematical difficulty appears to be one that must always surround any process involving important redistribution of matter under gravitational forces, and it is possible therefore that many of the key cosmogonical processes must for ever remain shrouded in this kind of mathematical uncertainty, and perhaps be capable of being established only by means of considerations of an indirect but circumstantial character. An instance of the kind of argument meant is provided by the differences of composition that exist between the four inner planets, and by their far smaller masses than those of the great planets. Both these features suggest that these terrestrial planets cannot have formed by the same process that produced the great planets, but there is the possibility that they originated from the central regions of the stream of material between separating components of the rotational break-up of a single primitive planet. Their combined mass is less than 1 per cent of that of Jupiter and therefore would represent an insignificant portion of the material involved in the main features of the disruption. Since the material would have already reached the liquid state in the primitive planet the difficulty concerning the formation of small bodies from gaseous matter at high temperature is avoided. Thus the development of the fission process in accordance with dynamical indications leads naturally to this possible mechanism for the formation of satellites, a process that itself arises naturally as part of the evolution of a primitive planet to a state of rotational stability.

Appendix

REFERENCES

References to all early publications are to be found in

Todhunter, *History of the Mathematical Theories of Attraction*, 2 volumes, 1873.

The principal early papers in the present connection were:

Meyer, *Crelle*, xxiv, 1842.

Riemann, *Gött. Abh.*, ix, 3, 1860; and *Werke*, p. 168.

Poisson, *Connaissance des Temps* for 1837 (published 1834). (This appears to contain the first reference to Jacobi's result that ellipsoids are possible equilibrium forms; stated by Jacobi in a letter to the French Academy 1834.)

Jacobi, *Acad. des Sciences*, 1834.

Liouville, *Journal de l'École Polytech.*, xiv, 290, 1834; and *Liouville's Journal*, xvi, 241.

Ivory, *Phil. Trans.*, Part I, 57, 1838.

de Pontécoulant, *Systeme du Monde*, ii.

A list of further early papers is given by Darwin, *Scientific Papers*, vol. iii, p. 119, taken from a report to the British Association, 1882, by W. M. Hicks, and a further list by Matthiessen, *Schriften der Univ. zu Kiel*, vi, 1859. Extensive references are also given by S. Oppenheim, 'Die Theorie der Gleichgewichtsfiguren der Himmelskörper', *Encyklopädie der Mathematischen Wissenschaften*, vi, 2, B, pp. 1–79; and by P. Appell, *Mécanique Rationelle*, vol. iv, 1921.

The problem of the stability of the forms was first broached in:

Poincaré, *Acta Math.*, vii, 259, 1885.

Further discussions of stability are given by:

Schwarzschild, *Ann. d. Münchener Sternwarte*, iii, 233, 1897.

Thomson and Tait, *Natural Philosophy*, 1879.

Routh, *Stability of Steady Motion*, 1877.

Hilbert, *Gött. Nach.*, 49, 1904.

Other papers by Poincaré include:

Phil. Trans., **198** A, 333, 1902.

Bulletin Astron., ii, 117, 1885.

The papers of Darwin are most readily accessible in:

Darwin, *Scientific Papers*, vol. iii, Cambridge, 1910.

Here errors in the original papers are corrected, and numerous detailed references to other earlier work and his own papers given.

Jeans's papers are as follows:

Phil. Trans., **200** A, 67, 1902. (In this paper it is concluded that the two-dimensional pear-shaped series is stable.)

Phil. Trans., **217** A, 7, 1916. (This concerns the three-dimensional pear, but in passing, the incorrect conclusion of the 1902 paper is referred to and corrected, p. 28.)

Shortened versions of these papers are given by Jeans in his two books:

Problems of Cosmogony and Stellar Dynamics, Cambridge, 1919.

Astronomy and Cosmogony, Cambridge, 1929.

Further accounts of different aspects of the problem may be found in the following:

Bryan, G. H., *Proc. Roy. Soc.*, **180** A, 187, 1888.
Love, A. E. H., *Phil. Mag.* (5), xxvii, 254, 1889.
Tisserand, *Mécanique Céleste*, ii, 1891.
Neumann, *Hydrodynamische Untersuchungen*, Leipzig, 1883.
Bassett, *Proc. Camb. Phil. Soc.*, viii, 23, 1892; and *Hydrodynamics*, 367, Cambridge, 1890.
Liouville, *J. de Math.* (5), iii, 1897.
Poincaré, *Figures d'Équilibre d'une Masse Fluide*, Paris, 1902.
Hargreaves, *Trans. Camb. Phil. Soc.*, xxii, No. 5, 61, 1914.
Cartan, E., *Bull. Sc. Math.*, t. 46, 332, 1922;
         Proc. Int. Math. Congress at Toronto 1924, ii, 9, 1928.

Liapounoff's principal papers are the following:

'Sur la stabilité des figures ellipsoidales d'équilibre d'un liquide animé d'un mouvement de rotation', *Annales de la Faculté des Sciences de l'Université de Toulouse*, 2e serie, t. vi, 1904; and t. ix, 1908.
'Sur un problème de Tchebychef', *Memoires de l'Académie Impériale des Sciences de St. Pétersbourg*, viii serie, xvii, No. 3, 1905.
'Problème de minimum dans une question de stabilité des figures d'équilibre d'une masse fluide en rotation', *Memoires de l'Académie Impériale des Sciences de St. Pétersbourg*, viii serie, xxii, No. 5, 1908.
Annales Scientifiques de l'École Normale Superieure, t. 26, 3e serie, 1909.
'Sur les figures d'équilibre peu différentes des ellipsoides d'une masse liquide homogène douée d'un mouvement de rotation', *Académie Impériale des Sciences, St. Pétersbourg* (in four parts: i—1906, ii—1909, iii—1912, iv—1914).

A statement of Tchebychef's problem is given by A. J. Pressland in a belated obituary notice of Liapounoff (1857–1918), *Nature*, **128**, 138, 1931, as follows: 'It is known that for certain values of the angular velocity the ellipsoidal form no longer serves as a surface of equilibrium for rotating homogeneous fluids. Does it change into some new form of equilibrium, which for small increments of angular velocity (momentum?) differs but little from an ellipsoid?' See also an earlier obituary notice, Stekloff, *Bull. de l'Acad. des Sciences de Russie*, 367, 1919.

Other more recent publications discussing particular points are:

Baker, H. F., *Proc. Camb. Phil. Soc.*, xx, 181 and 198, 1920; xxiii, 1, 1926.
Humbert, P., *Compt. Rend.*, **170**, 38, 1920.
Lichtenstein, L., *Math. Zeit.*, i, 228, 1918; and vii, 132, 1920; and *Gleichgewichtsfiguren rotierender Flüssigkeiten*, Springer, 1933.
Lamb, H., *Hydrodynamics*, Chap. xii, 6th edn., Cambridge, 1932.
Mikhailenko, B. G., *Thèse*, Paris, Gauthier-Villars, 3e part, 60, 1932.
Appell, *Bull. Sc. Math.*, t. 45, 10, 1921; and *Compt. Rend.*, **171**, 761, 1920.

Accounts of the theory of ellipsoidal harmonic analysis and references to the earlier work on Lamé functions may be found in the following:

Poincaré, *Figures d'Équilibre*, 1902.
Whittaker and Watson, *Modern Analysis*, Chap. xxiii, Cambridge, 1946.
Hobson, *Spherical and Ellipsoidal Harmonics*, Cambridge, 1931.
Appell, *Mécanique Rationnelle*, vol. iv, 1921. (Gives comprehensive references, pp. 289–294.)

Accounts of the cosmogonical implications associated with the problem are to be found in the following, which also contain many further detailed references:

Poincaré, *Hypothèses Cosmogonique*, Paris, 1911.
Darwin, *Scientific Papers*, Cambridge, 1910.
Jeans, *Problems of Cosmogony*; and *Astronomy and Cosmogony* (see above).
Jeffreys, *The Earth*, Cambridge, 1924 and 1929.
Russell, H. N., *The Solar System and its Origin*, New York, 1935.

Darwin's calculations on the limiting forms of close double systems are to be found in:

Scientific Papers, iii, 436–524, Cambridge, 1910.

The values of the angular velocities of Table VI are taken from

Jeans, *Problems of Cosmogony,* p. 63.

Diagrams of some of the limiting forms may be found in both these works (Darwin, pp. 508, 509; Jeans, p. 64).

References to other papers on the cosmogonical aspects of the problem of rotating liquids may be found from the General Index of the Monthly Notices of the Royal Astronomical Society.

INDEX

For EU product safety concerns, contact us at Calle de José Abascal, 56–1°, 28003 Madrid, Spain or eugpsr@cambridge.org.

www.ingramcontent.com/pod-product-compliance
Ingram Content Group UK Ltd.
Pitfield, Milton Keynes, MK11 3LW, UK
UKHW060313090126
466816UK00021B/475